王劲韬 著

景观规划徒手表达
Hand Drawing Presentation of Landscape Planning

By Wang Jintao

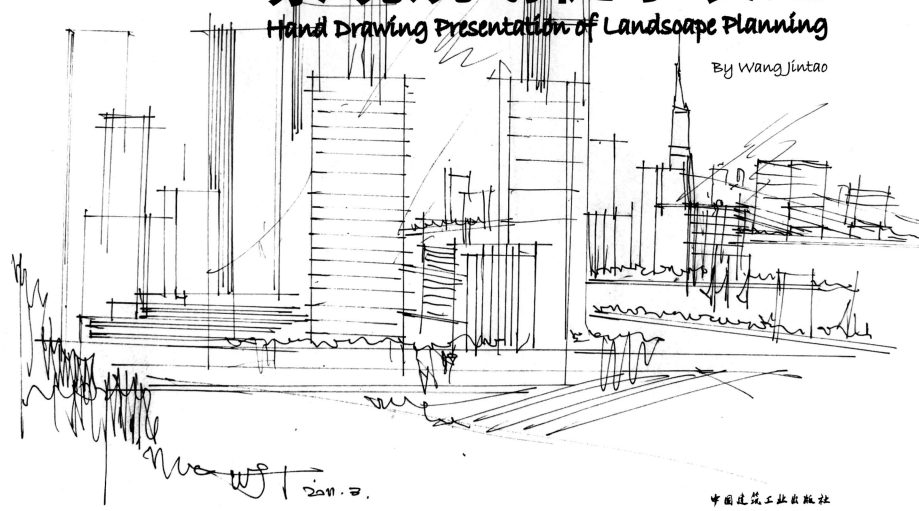

中国建筑工业出版社

图书在版编目（CIP）数据

景观规划徒手表达 / 王劲韬著. —北京：中国建筑工业出版社，2012.4
ISBN 978-7-112-14100-5

Ⅰ. ①景… Ⅱ. ①王… Ⅲ. ①景观规划—景观设计—绘画技法 Ⅳ. ①TU986.2

中国版本图书馆CIP数据核字(2012)第035800号

责任编辑：常　燕
版面设计：叶　飚

景观规划徒手表达
王劲韬　著

*

中国建筑工业出版社出版、发行（北京西郊百万庄）
各地新华书店、建筑书店经销
广州市弘志广告设计有限公司制版
广州市一丰印刷有限公司印刷

*

开本：787×1092毫米　1/16　印张：12$^7/_8$　字数：313千字
2012年6月第一版　2012年6月第一次印刷
定价：**78.00**元
ISBN 978-7-112-14100-5
　　（22140）

版权所有　翻印必究
如有印装质量问题，可寄本社退换
（邮政编码 100037）

谨将此书献给

热爱景观手绘的学子们、设计师们

你们的鼓励和持久的参与，使手绘得以重新回归设计师的职业视野

并作为设计的灵感之源被发扬光大

本书受"中央高校基础科研业务费专项资金资助"（项目编号：TD2011—31）
Supported by "the Fundamental Research Funds for the Central University"（No.TD2011—31）

序言 / PREFACE

作为说明和交流工具的手绘图

Hand drawing as a tool of illustration and communication

手绘图是设计师自己的交流工具，也是设计师与业主的交流工具。建筑画作为建筑设计发展史上的一个阶段性的成果，用以评价或者阐述一些观点，这一手法恐怕是从文艺复兴就开始了。比如达·芬奇手稿中列举的那些穹顶、梁托、著名的维特鲁威人图示等，达·芬奇手稿中的那些草图似乎更多的是作者与自己的思想做交流的过程展示，通过草图描述并记录自己的思考轨迹，所以不如称之为笔记。而伯拉孟特为圣彼得教堂所作的那些精粹的平面和立面草图则带有更多地向甲方业主（教皇克莱门七世）阐述其空间理想的意味。

在景观建筑表现图的历史上还有一类纯粹理想化的图纸，这是独立于以上两类的特例。诸如英雄主义建筑师艾蒂安·布雷的大都市教堂和牛顿祠，以及浪漫主义大师辛克尔早年的一些舞台设计，如他为沃夫冈·莫扎特的歌剧《魔笛》所作的极富象征意义和历史主义隐喻的舞台场景。这种浪漫主义的艺术追求到了20世纪，在渲染大师沙勒尔的许多方案表现图中仍然得到了充分体现。直至近代美国人赖特的作品——英里大厦的渲染图，似乎都是为阐述一种理想，和对未来的一种预言。这类作品更类似于今日的动漫设计，诸如宫崎骏的动画，抑或技术理性下的变形金刚等形象。电脑软件的场景渲染技术在这一方向上所展示的超能力，已经使这类作品完全不需要手工的干预（除了最初的基本形），其作品感染力也大大超过了当年辛克尔创作《魔笛》一类场景所能达到的水平。

Hand drawing, not only serves as a tool of communication for a landscape designer him or herself, but also a tool of communication with clients. Architectural drawing is a periodic achievement in the history of architecture design, which used to comment or depict ideas. Perhaps this technique can be traced back to the Renaissance. From the corbel and vaults to the famous Uomo Vitruviano, the drawings of Leo da Vinci are more like an illustration of the process that the drawer communicated with himself. Drawing was used to describe and record the tracks of his thoughts, thus rather a note than a drawing for him. As to the precisely drawn plan and elevation sketches of the Basilica di San Pietro in Vaticano by Donato Bramante, they are more inclined to depict space ideal for the client (Pope Clement VII).

In the history of landscape architecture presentation drawing, there is another kind of drawing-the pure idealized drawing which distinct from those two kinds. Such kind includes Metropolitan Cathedral and Newton Memorial of Heroism architect Etienne Louis Boullee, and some stage designs of Romanticism master Karl Friedrich Schinkel in his early years, e.g. the stage scenes he designed for Wolfgang Amadeus Mozart's opera "The Magic Flute", which are full of symbolic and historical connotations. Even in the 20th century, the artistic pursuit of Romanticism were still found overwhelming in many plan presentation drawings of rendering master Schaller, let alone American architect Frank Lloyd Wright's The Mile high Illinois, whose rendering seems no more than depicting of an ideal and foreseeing the future. These works are more similar to cartoon design, such as Miyazaki Hayao's animation, or the figures like transformers in the context of rational techniques. In fact, the rendering techniques of computer softwares performed a presentation so great that this kind of works can be kept off any hand drawing (expect the basic form), and the final works are even

左图1：辛克尔他为莫扎特的歌剧《魔笛》所作的舞台场景设计，以古埃及神庙为背景，纪念碑式的构图，将场景的历史感、永恒性和异国情调熔于一炉，画面充满了想象的张力。

左图2：沙勒尔的历史题材表现图——罗马雕塑群，所有的形式——雕塑、券门、栏杆都化为历史感的符号，真正的决定因素是光和影产生的节奏和空气感，这是沙勒尔作品的一个共同特点。

left 1:Stage design by Schinkel for Mozart's opera The Magic Flute. The stage scene was set in the background of a temple in Ancient Egypt and has a monumental-type composition. These made the whole scene embody with historic, eternal and exotic atmosphere at the same time and have the tension to stimulate imaginations.

left 2:Expression of historic theme by Salter-Roman sculpture group. All the characters-sculptures, arches and handrails are put into historic symbol. The common crucial factor in his works is the air sense and thyme created by lights and shadows.

伊利诺伊摩天大楼：1956年计划在美国芝加哥建造。这座摩天大楼的高度设计为1609m，设计师是弗兰克·劳埃德·赖特，他认为当时提议建造这样的宏伟建筑是可能实现的。这座摩天大楼包括528层，总面积达到1714990m²。伊利诺伊摩天大楼方案提出后发现一系列问题，比如：维护电梯的空间将占据较低楼层的所有可用空间等，因此，这项建造摩天大楼的计划被迫夭折。

所以说，今日手绘，撇开简单的技法优劣方面与电脑的对比，其发展方向已从过去追求尽善尽美的光影表达和环境渲染，转向另一种境界的追求：对创作意向尽可能迅速地把握、抓取和记录，进而推向一种表达瞬间意念的方向。事实上，许多稍纵即逝的设计构思唯有通过快速手绘图，才有可能保存和记录下来，这种捕捉瞬间意念的能力是在电脑前不可能实现的。清华大学的朱文一教授曾就第一届建筑手绘图大赛撰文提出一种特殊的手绘工具——手鼠笔。这是一种将鼠标和传统画笔完美结合的超一流的快速设计工具，可以在电脑屏幕前高速模拟钢笔、铅笔、水彩、色粉甚至油画等各种机理和特效。但实质问题是，这一切特效是来自于作者的手绘创作，还是手脑配合的产物，只是高科技可以使我们站在更高的平台上，用比传统绘画多上几千倍的色系选择进行创作。但究其根本，仍然没有否定手工绘图在设计，尤其是在创作阶段的重要价值。朱文一先生用幽默的语言表达了目前作为中流砥柱的一代中青年建筑师对

much more appealing than such scene as Schinkel designed for "The Magic Flute".

The Mile high Illinois, designed by Wright, was planned to be built in Chicago in 1956. With a total height of 1609m, the skyscraper has 528 floors, and the total area amounts to 1714990m². However, a series of problems rose afterwards, for instance, all the available space of the lower floors had to be used for elevator maintenance. Therefore, the project had to be terminated.

So I think that hand-drawing (despite simple technical disadvantage vs. computer), instead of pursuing perfect light-shadow expression and scenery rendering, are heading to the pursuit of another stage nowadays: to seize, grasp and record the inspiration as quickly as possible, to present the flowing ideas in a twinkle. In fact, only the quick sketch can keep a record of many instant ideas, which cannot be realized by computer. In his essay on the 1st Architectural Hand drawing Contest, Mr. Zhu Wenyi, Professor of Tsinghua University, put forward his ideal drawing tool - "hand-mouse pen", an imaginary quick design tool which combines mouse and traditional paintbrush perfectly and can simulate the mechanism and special effects of pen, pencil, watercolor, gouache, and even painting. It is only that High-tech raises us up to a higher platform where creating with thousands more colors than traditional drawing is available. However, all those special effects still originate from the illustrator's hand drawing creation and come from his hand-brain cooperation, the essential part of this imaginary tool never denied the important role that hand drawing played in design, and especially in the creation process. Mr. Zhu Wenyi's humorous language expressed a mainstream view of the present mainstay young and middle-aged architects on the relationship between the inheritance of traditional drawing and the application of modern computer rendering techniques: to combine the computer

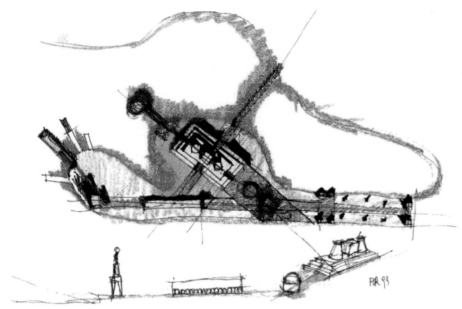

阿尔多·罗西的作品——韩国某旅馆　A hotel in Korea designed by Aldo Rossi

阿尔多·罗西的作品——科学公园　Science Park designed by Aldo Rossi

于传统手绘继承和现代电脑渲染技术应用这两者关系的一种主流观点：将电脑技术与传统手绘相结合，更好地抓住设计师的灵光一现。

此外，在艺术化地表现创作思维领域，手绘图同样具有不可替代的优势，这也是手绘图作为交流工具的一个重要特征。我们在手绘表现图里所呈现的笔法形式、主观情趣等内容往往更多的是在传达设计者在设计品位、设计兴趣和关注点等方面的信息，而不是亦步亦趋的实景模拟，这往往是解读这些手绘"天书"的前提。这就有点像达·芬奇笔记中的那些看似潦草的说明图，在承认作者不凡的艺术灵气和激情的同时，更需要解读作者用隐晦的方式传达出的设计信息和趣味焦点。在纯建筑领域，阿尔多·罗西是这方面的典范，他的建筑草图在我看来是用孩子一般的天真烂漫表达出严谨缜密的建筑思维。作为一种独立的艺术表达形式，大师的手绘草图往往寥寥数笔就能表达出意味无穷的境界，因为这寥寥数笔包含了作者深厚的学识素养和对设计项目深刻的理解，这就是所谓的功夫在画外。这应该是建筑手绘图在自我交流和团队交流方面的第二个发展方向。

techniques with traditional drawing, so as to catch the instant inspiration in design. And that's where hand drawing may head to.

Besides, in the field of artistically presenting creation thinking - also another important feature of hand drawing as a communication tool - hand drawing shows its irreplaceable advantage. The style, form, and personal temperament presented in a drawing is more a delivery of the information about the illustrator's taste, interest, and concerning points in design than the rigid simulation of realistic view. And this often serves as a premise to interpret these wordless works. Just like the seemingly hasty illustrations in Da Vinci's notes. Besides acknowledging his outstanding art ingenuity and passion, it counts more to know the referring information and focus of interest which are delivered in an obscure way. In the field of pure architecture, Aldo Rossi is a role model. His architectural sketch, from my point of view, though seemed as simple and innocent as a child, contained strict and meticulous architectural thoughts. Hand drawing as an independent artistic expression, can often express endless meaning in several simple strokes by great draftsmen, for these strokes contained their rich academic possession and deep insights on the designing project. This is what we call "art of drawing coming from the outside". This should be the second development direction of architectural hand drawing with respect to inside and outside communication.

左图：彼得沃克事务所的草图、草模、成图和正式放大模型，一个从创意到手工图形、工作模型和式成图的完整创作过程。

left: The sketch, preliminary model, final draft and formal magnified model of Peter Walker and Partners-a complete creative process from an idea to hand drawing, model and final draft.

作为设计思维过程的展现和具体化的手绘图

Hand drawing as a specified presentation of designing process

设计过程是一个思考的过程，也是一个寻找最合适表达途径的过程。在这一过程中，电脑和手绘所承担的角色应该是有所区分的。电脑的优势在于表达最终成果，其光、色及环境氛围的逼真程度是手工渲染望尘莫及的，但是这仅止于最后阶段。电脑渲染技术无疑为景观建筑表现艺术开辟了更加广阔的前景，但当前的问题在于，电脑渲图技术在大获发展的同时，带来了设计过程中的各种"变异"，简言之即以"表现"代替"设计"。这种趋势发展下去无疑是相当可怕的：用软件研究代替空间研究，用数字堆砌代替对艺术品质的推敲，说明文本完全为各式"美图"所替代，进而，设计成果展示几乎异化为竞图比赛。于甲方而言，在专业视野有限的前提下，往往无法在创作初期就触及项目的实质内容，而直接被"美图"所俘获，这种"电脑美图大全"给项目带来的不仅是设计变异、变味，而且这种变了味的所谓"设计"往往根本无法落地，成了空有一身好皮囊的花瓶和空中楼阁，这对项目的贻害是显而易见的。而对设计主体——设计师，或是那些即将成为设计师的学生而言，这种影响就更大。对景观设计过程的推敲远不如表达图经得起考量，"构思"一词更像是说"构图"，我看到越来越多的学生设计（包括纯研究性的竞赛设计）往往只有成图、电脑模型，而没有一张过程草图，这与设计的科学过程是极不相容的。

这种数字化"替代"、扼杀的可就远远不止那些设计的美感和

Designing is a process of thinking as well as a process of finding the best way of expression. However, the role computer plays in this process should be distinctive from hand drawing. Computer has its unique advantage in expressing the final result; the verisimilitude of light, color and environment are far more perfect than hand rendering, only in the final stage, though. No doubt computer rendering techniques opens a better future for landscape expression artistry. However, with its dramatic development, it brings multifarious "variations" -namely, expression takes the place of design. It would be much more terrible if keeps growing like this: software development takes place of space research; digital generation takes place of artistic consideration; illustrative text totally replaced by pretty pictures; finally, design presentation becomes a variant of picture competition. As to the clients, they are often attracted by pretty pictures soon since them unable to get the essential part of a project for their lack of specialized point of view. Unfortunately, those "pretty digital picture collection" brings no more than an abnormal variant of design which "designed" nothing but good-looking castles in the air. Its deteriorating effect is apparent. As to the mainstay of design--the designers and the designer-to-be students, it could be even worse. Their weighing in the designing process becomes far less than their consideration in way of expression. For them, "designing" is much closer to "picture composition". I have seen more and more students' "designs" (pure researching designs for competition included) are only computer models and renderings but designing sketches. Rather than designing, they are organizing pictures, which is contradictory to the scientific process of design.

Perhaps more than sense of beauty and creation, even random changes and richness in design will be killed by the digital replacement. As a result, the students give up individual thinking when they

左图：吴良镛先生早年习作《巴黎圣母院》和《威尼斯圣马可广场》。老一辈建筑师的作品仍是今日学生的学习榜样。

left:"Cathedral Notre Dame de Paris" and "Plaza San Marco in Venice", drawn by Mr. Wu Liangyong in his early years. Senior architects' works are still examples for present students so far.

创意，还造成草图成型过程中的许多随机性变化的被扼杀和设计丰富性的缺失，最终导致学生在学习阶段就将发挥主观能力的权利拱手让出。随之而来的设计构思，分析，归纳能力以及最终创作能力的缺失，也就不足为怪了。这就是为什么我们的园林专业在录取分数逐年上升的情况下，学生素质、综合能力却急剧下降的原因之一。这一点，我们只需稍稍回顾一下我们的前辈大师们在学校期间的习作，两代建筑学子之间的差距就非常明了了。

艺术创作最重要的能力是想象力和感悟力。想象力的极致，正如布拉蒙特所创造的奇迹——将伟大的万神殿大穹顶"举起"，搁到圣彼得大教堂之上。而感悟力，我们则需要做到像景观设计前辈布朗先生那样，时刻感受到大地脉搏的跳动，时时刻刻能发现土地的潜质和风景的诗意。而单纯强调电脑渲染技术、忽视手绘推敲过程、忽视手脑并用的训练，这种设计模式无疑剥夺了设计师发挥想象和创造的权力，感悟的过程更是被直接简化而近于虚无！但如果设计师被剪断想象的翅膀，他还能有什么可为之处？电脑渲染的优势在于最真实地再现自然，但也仅仅是重现视觉反馈，而想象则足以使人超越视觉的简单存在，用设计师的经验和智慧抽取场所的本质，创造超越简单存在的视觉自然，我们称之为灵光一现，而画出来的灵光一现更是相当可贵的，因为设计需要的正是这种在对普遍存在物的感悟和超越的基础上，进而再创造的过程。

are still learning; no surprise that they are followed by losing conceiving, analyzing, and inductive abilities and finally creativity. That's why the quality and comprehensive ability of our Chinese landscape majored students has been declining year by year, as a contrast to the rising admission line of the major. Just glimpse back at the works our predecessors have done in their school, you'll see the gap between the two generations of architects.

Imagination and perception are the two most important elements in artistic creation. An extreme example of the former is the "miracle" Donato Bramante has created: he "raised" the vault of Pantheon and placed it onto Basilica di San Pietro. As to the latter, we can learn from landscape design predecessor Joseph E. Brown, who can always feel the impulse of the earth, find the potential of land, and see the poetry of scenery. The design mode that simply emphasizes the digital rendering techniques, while ignoring the process of elaborate designing and the training of hand-mind cooperation, is no less than depriving the imagination and creation from a designer. What a designer could do if he was cut the wings of imagination? Undeniable, computer has its advantage in representation of the nature, but it only focuses on the visual feedback. But imagination works go beyond visual sense. A designer with experience and wisdom can extract the essence of a place and create a visual nature that beyond simple existence - what we called inspiration in a twinkle. Since to perceive into the universal existence and to create beyond it is what design requires. It would be much cherished if the inspiration can be drawn out.

So here is my conclusion: even in the times of digital high-tech development, it would be a terrible mistake to abandon the weighing process of sketch, hand drawing and preliminary model. Not only because preliminary designing by computer (almost unrealizable actually) cannot seize the flowing

我的结论很明显，即使在数字化技术高度发达的今天，主动放弃手绘图、草图、草图模型的推敲过程也是不足取的。这不仅是因为直接采用电脑进行初步设计（实际上几乎做不到）对于许多灵光一现的创意根本无法把握，更因为艺术创作的基本规律使我们相信：为人所服务的空间，与文化紧密相连的表现艺术，不可能也不应该被简化为一种流水线式的"作品"。更何况在这种数字化过程中，许多细节在剥离了设计本身的模糊性、可变性特质以后，便会显得空洞而多有掣肘，许多成熟设计师是深悟其中三昧的。

具体而言，手绘图的优势表现在以下几个方面

1. 在于方便快捷地把握一些转瞬即逝的创意、符号，并通过后续的详细设计对创作意图加以完善、说明。这种方式类似于许多成熟的建筑师在餐巾纸、烟盒上所作的天书式的"草图作品"。在专业领域中，日本人安藤忠雄是这方面的榜样，安藤忠雄的许多方案是随机而来的，几乎是想到什么好点子就会及时画下来。从速写本到餐巾纸、报纸，几乎任何时候有灵感都能抓住，并以自己的理解加以表达。这类似于天书的表达，虽然只有设计者本人能够读懂，但却足以起到记录、提示的作用，是下一步深化的重要基础。安藤忠雄的草图的意义恐怕不仅在于记录，这种绘图是情感跳跃过程的流露，据此几乎能听到建筑师的心声，生命的能力所倾注的图像，绝不仅是数字化生产线上能实现的。

inspiration, but also the basic principle of artistic creation makes we believe that neither the space serving people nor the expression art that closely related with culture, could or should be simplified as an assembly line produced "works"; let alone the creating activity, being stripped off ambiguity and variability in the digitizing process, becoming empty and limited. And that had been deeply understood by many experienced designers.

The advantages of hand drawing are specialized in the following aspects:
1. It is convenient for designers to grasp the twinkling creative ideas and symbols quickly, which can later be perfected and demonstrated by the following detail designs. It is particularly similar to many experienced architects' obscure "sketch works" which were drafted on napkins or cigarette packets. In specialized field, Ando Tadao is an example. Many of his schemes seem came by accident. Actually, he has a habit of drawing out any good idea that occurs to his mind any time on any materials - napkin, newspaper, etc. - and presenting the inspirations in his own way. Though this kind of presentation can only be read by the illustrator himself, it has fulfilled the role of records and reminders and laid an important foundation for the deepening step. Perhaps more than recording, Ando's sketches reflect the leaps of emotion that we can almost hear the architect's heart. Those heart-made pictures can never be produced by digital assembly lines.

In fact, this "napkin sketch" cannot be emphasized more any time than in modern designing process. Ando Tadao holds the view that just some elements and specified scales added, these sketches can be modified into "design presentations". Neither original nor determinative, many the following works are often the refining and specifying process. But what really matters twinkled on common napkins - the inspiration.

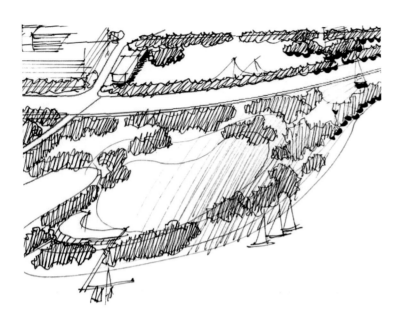

事实上，现代设计过程比以往任何时候都少不了这种"餐巾纸式"草图。安藤忠雄认为：给这些草图上加上一些要素和具体尺寸后，就能整理成"设计画面"，后续许多工作往往属于完善和具体化过程，并不具有原创意义，决定性之举往往是不起眼的餐巾纸上的"灵光一现"。

2. 草图有利于促进设计思考和完形。如果我们承认设计是一个既有灵感的凸现，又有逐步完善的过程，那么草图在这一过程中则是思考的工具。从这个角度上讲，设计不如说是"画"出来的，设计制图与纯艺术的绘画不同点在于后者是成竹在胸，落笔在后，设计则往往是从一个快速闪现的灵感起步，通过逐步深入的思考而渐渐清晰的构思过程，草图则有利于促进这一思考和完形的过程。所以从某种意义上说，设计是"画"出来的。从大的概念性调整过程到每一个设计细节，往往都是在不停顿的画的过程中逐步完成的。不仅如此，新的灵感也会在草图完善的过程中逐步闪现出来，这方面个人的体会是颇为深刻的。如果缺少"画"的过程，我们的设计难免会像个干瘪的老太太。反之，业内很多艺术绘画和平面设计好手跻身于景观设计，并有所斩获，靠的也是这种手头硬功夫。

3. 作为说明性的草图更容易突显设计师的主观意念对元素、符号的概括、提炼作用，即更能抓住有目的选择和强化某些重点部分，

2. It helps the thinking and completing of the design. If design is a process involved both inspiration highlighting and gradual completing, sketch is a tool of thinking in this process. In this sense, designing is almost drafting. Different from pure artistic drawing being drawn out as an already existed image in the illustrator's mind, design usual starts with a flashing inspiration, and becomes clearer and clearer with deeper thinking. So in a way, designers "drafting" designs - from conceptual adjustment to every details designing in the process, are often made by constant drafting, which helps the thinking and completing. Moreover, usually new inspiration comes in the completing process, for which I had experienced deeply. Leaving out drafting, our designs can never be exempted from boring. An opposite aspiring story is that when many good hands of artistic painting and graphic design turn to landscape field, they usually succeed and lead fruitful careers, for their great proficiency in drafting.

3. It highlights the designers' individual epitomizing and abstracting on elements, symbols, etc. The illustrative sketch, with better control of the oriented adoption and emphasize of certain parts, can deliver its message more selective and subjective, and then in a more personal style of being more directive in design. In a word, hand drafting are more likely heading to a style of clearer elements, simpler frame, and more explicit presentation. Particularly in the sketching phase, this kind of ideas-contained expression of personal style works would make the picture (design) be more recognizable and more readable. Many masters' works can be recognized in a glimpse and their presentations are so clear that even the highly developed computers can't achieve.

右图：汉德森的一个案例（星期五晚上：菩提岛）
right:an illustration of Henderson（FRIDAY EVENING:PUTI ISLAND）

这组草图是从本人的老师和合作伙伴，汉德森教授的速写本上节选的一个案例，简素明确的画面传达出本人对场地的构造、特征和意境的层层深入解读过程，以及高度个性化、习惯性的诠释方式。

画面传递的信息更具有选择性、主观性，进而具有个性化艺术风格以及设计上的指向性。手绘图更容易导向一种元素更清晰，画面更简略，表达意图更明确的图式风格。尤其在工作草图阶段，这种个人风格和作品的意念化表达会使画面（设计）显示出极强的可识别性乃至可读性。许多大师的作品往往能在瞬间被辨别出来，其清晰程度令现代功能极强大的电脑也望尘莫及。

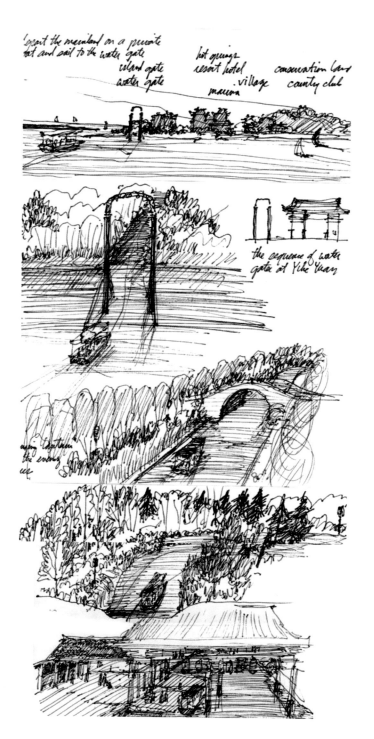

八点半：从北京出发开车三个小时之后。车子停在岸边。我们从菩提岛上的码头上登船。这是个精悄的温暖的晚上。越身的南侧地也高比较小。我们可以闻到空气中咸味的味道，听到海鸥的叫声。船上还有大约二十个人和一起，我们悄悄地驶入港湾。缓缓走过狭窄的通道。两岸的帆船和小艇花桨并然有序地停在停泊处。我看到岛上的水门出现在我们面前。

八点四十分：我们穿过水门之后船行进至一个狭窄的水道。这水门使我想起了古代中国纪念伟大航线的牌楼，或者在日本宫岛所见的水国社。尽管这水门与颐和园挂凌霄楼太后上岸的码头上那个高面优美的牌楼更为相似。水面上荡漾的波纹倒映出红色的塔。柔和月光环绕着金色形的拱门。

八点四十二分：运河仅足够两艘船擦身通行。但我们并没有遇到另外的船。岸边的植物退开海水沿着海边的沙滩丛生长着。水边稍的地方长着修剪整齐的芒草坪，睡莲和芒花芦苇。精选的水边，挖叶树，蓝芦，月桂树，玫瑰都从后面更高处浓密的松树，橡胶树，黄柠树下面生出来。海水的清新和海边玫瑰的香气浓郁在空气中，我深深地呼吸，开始感受这个岛屿的表面。

八点四十四分：我们穿过一座木制的拱桥，桥上还有朦胧的白色灯笼。像一个小小的月亮。桥上并没有人。

八点四十五分：我们跟着运河转过一个弯，而后看到了面更宽的船只。在左边，月光朗朗的天空映照岛上的桂枝花，酒店的轮廓从其中逐渐升高。我们渐渐听到远处的欢声笑语。

八点四十八分：我们的船停靠在一个搭建得伸入水中的宽宽敞的码头。我们到达了酒店。水边的速草里也是群一样的白色灯笼。身着深红色长袍的礼宾欢迎我们的到达，将来客们的行李筐上行李架。我们步行经过走廊，穿过这个度假村地入口的大木门。

手绘图的未来

数字科技、电脑模拟软件使图面的精度达到了传统绘图方法无法望其项背的程度。当代设计师的工作台面上总是少不了各种各样的数字化模拟仪器，作为后期表现的主力，这种人机协作的方式和技术支持毫无疑问将为现代设计提供更广阔的空间，但这并不意味着草图纸、绘图桌就此退出设计过程。我的观点非常明确，在设计构思的最初阶段进行全电脑生产线式的制作是不经济、不现实、不值得提倡的。为工作草图建立数字化模型不仅耗费大量的人力，而且时间成本极大。电脑图纸更适合的角色是作为一个终端产品，其间对于一些有价值的草图方案可以适度发展成为数字化草模，草模和草图在设计初期可以并行发展，但摈弃手工草图，完全依赖数字模型必然会限制多种方案的并行发展。而且，电脑渲染图这个终端产品的质量和艺术性还是与设计操作者本人的艺术素养和制图水平紧密联系，而不可能完全作为一种纯智能化的产品独立存在于电脑软件中。事实上，高度发达的现代设计过程比以往任何时候更需要"餐巾纸"式的草图。

但无可否认的事实是，我们当前的建筑景观表达正逐步走向追求单一目标的，追求尽善尽美的环境渲染和绚丽的画面效果的终端式产品，而不是作为一种思考的辅助性工具。官方的设计竞标在相当程度上引导了这一发展方向，对方案的评审和最终取舍往往依据图面的表达效果，却很少关注于过程性创作，许多精彩的过程性草

The future of hand drawing

On the other hand, digital techniques and computer simulation software has raised the picture rendering so high that traditional drawing could never exceed. We can always find various digital simulation instruments on the workbench of contemporary designers. As computer serves as a major force in the final presentation, no doubt that human-machine cooperation and technical support will create a wider stage for modern design, but it doesn't mean the sketching paper and drawing table will leave the design process. So my point is, it's less economic and unrealistic to adopt the computer workshop-producing in the very beginning of design, and thus unworthy to be tried. (Actually, to create digital model for sketches not only makes people exhausted, but also costs a lot of time) I think it would be more appropriate for computer drawing to act as a final product. Some valuable sketches could be developed into digital preliminary models, or both in a paralleled way of developing; however, if hand drawing were abandoned, the overall development would be restrained by relying on digital model totally. Anyway, computer renderings the final product, whose quality and artistry still close link with the artistic quality and drafting capacity of the designing operator, would never exist in the software independently as a total artificial intelligent product. In fact, the highly developed modern design process requires the "napkin sketch" more than any time.

But the undeniable truth is our present architecture presentation becomes an end product which pursues a single result-environmental rendering and gorgeous visual effect as perfect as possible-rather than a thinking-aids aided tool that it should be. To a large extent the official competitive biddings of design has led the way of this abnormal trend. The bidding reviewers are often more attracted by the visual effect of presentations and less concern about the progressive creation, thus many wonderful progressive sketches and preliminary

图,工作草模被那些渲染精致、热闹非凡的电脑表现图所掩盖。这让人想起半个世纪以前,柯布西耶的那些精彩的研究性草图在官方竞图比赛中被直接废弃的悲哀经历。同样,在当代中国设计一线,这样让人啼笑皆非的闹剧天天都在上演。由于欣赏能力和艺术、技术素养方面的限制,由于经济政治等方面的驱动,在我们这个领域,草图这一具有决定意义的工作步骤正逐步淡出我们的专业视野,而手绘图这一景观建筑师必备的基础能力也为年青一代设计师所忽视,这种方向上的偏差以及由此对这个行业未来发展带来的消极影响也是不言而喻的。

我们今天所关注的正是竞图大赛中为官方所忽略的那些柯布西耶式的间断性图纸及其表述语言。其中大体涉及到题材的取舍,元素提炼和个性化的表达方式,以及将各种元素权衡、综合之后形成的设计师自己的语言和阶段性成果的表达方式。我认为,这种概括性语言和表达能力的提炼,对于个性设计风格的形成和完善深具意义。同样在大规模项目团队协作中,这种能力也可以成倍提高各级交流的效率。

本书所选作品以彩铅和马克笔混合制作为多,极少数早期作品中使用过水彩。因为在强调快速、达意的交流性图纸中,马克笔、彩铅有其天然的优势,我主张大力推广这种画法。为了达到这一目的,

models fade before those exquisite glamorous computer renderings. It reminds us the pitiful experience of Le Corbusier half a century ago, when his wonderful researching drawings were directly rejected in the government's bidding of "pictures". Unfortunately, this awkward and embarrassing situation in designing field is occurring everyday in contemporary China. Because of the limitation of appreciative intelligence, artistic and technical knowledge, and the drive of economic or political considerations, hand drawing, as a determinative designing process, is fading away from our specialized field. And this indispensable basic skill is also ignored by the young architects. It is an obvious developing deviation and it brings negative effect to the future development of our industry.

So what we are concerning about nowadays is the Corbusier-style discontinuous drawing and its expression which was ignored by government reviewers in design bidding. It mainly involves the choosing of subject matter, the extracting of elements, and the personal way of expression: an expressive language and way of expressing phased achievements of the designer's own after considering and integrating various elements. I think the refining of recapitulative language and expressive ability not only plays a very important role in the forming of personal style of design, but also helps to multiply the efficiency of communication among all sectors in large-scale project cooperation.

The works selected in this book are mainly rendered by color pencils and markers (rare early works used gouache), for I strongly recommend taking the natural advantage of color pencils and markers in communicative drawings in which quick and expressive are emphasized. For this purpose, I presented textual description for the illustrations and works in this book. I wish to share my years' experience in landscape drawing and some simple but practical skills with the readers, and give some

我在书中的例释和作品集等部分以文字说明的形式,总结了多年景观绘图的经验,以及一些粗浅易学且相当实用的做法惯例,以期与读者分享,并就快速作图技能的提高给广大学生、考生和初级设计师提供一些参考。

需要说明的是,书中有些极快速条件下的图例,快到10分钟成图。在构图、用线等方面均未加以考量,更类似色彩速写。为的也是适应考研中的快速方案表达的需要。在今年的几期教学录像中,我也做过类似的示范,读者的评价褒贬不一,大致是认识角度差异所造成的。但我想指出的一点是,这种快速训练对准备考研和入职考试的设计师非常有必要。就像美院的学生一小时交十张"高速"速写图一样,实质是要求设计师瞬间把握主旨,既是练手头也是练眼力,对设计水平的提高是大有裨益的。这一过程应该尽量快捷,而且绘制过程应该不拘一格,可以使用各种各样触手可得的材料,快速成图,可以快得像天书,像安藤忠雄那样,也可以稍微细致,以便于团队交流或作为与甲方交流的图像参考,但作为工作草图的设计表达总的原则是快!这种思考性创作有时会快到难以想象,由于技巧纯熟,作者得全神贯注于方案本身,而无须顾及表达技巧,最佳状态是心手合一,设计场景在思考过程中似乎能够自动生成;这种创作往往具有高度概括的真实性,完全摒弃了照相式的真实记录,只记下作者感兴趣的作用于设计思维的那些要点,与设计无关

clues of improving quick sketching skills for the students, examnees and junior designers.

It should be noted that some speed sketch illustrations in this book didn't spare too much consideration on framing or applying of lines and are more of color sketching, but seek a quick expression that can be completed in 10 minutes, which is also in accordance with the demands of graduate admission examination. I also made similar demonstration in some teaching videos this year, but received praise and depreciating mixed feedback from the audience due to different points of view. Nonetheless, I wish to point out the necessity of quick skill training for the examinees for postgraduate and designer applicants. Just like art school students are asked to hand in more than 10 "speed" sketches in an hour, this requirement, essentially, is to seize the subject in a minute by insight and drafting skills training, which helps to improve the design. So this process should be done as quickly as possible yet the operation can be as free as possible: use any materials available to finish a draft. No matter too hasty or slightly delicate as Ando's works for better inside communication or outside reference, the general principle of this sketch as design presentation is to be "quick"! Sometimes this creation from thinking is too quick to imagine, but a skilled illustrator would devote himself into the schema and ignore the expressive skills, and almost automatically draft out the designed scenery in his mind when in a perfect state. Because of its highly generalized truthness, this kind of sketches completely abandoned photographic record of reality. Instead, it takes down what the illustrator is interested while ignoring all the details that are unrelated with the design. Being so quick and "accurate" and with stronger art appeal, hand drawings play a determinative role in the design, and this phased task can never be done by computers.

的细节一概省略，所以它才既快又"准"（切题），而且更有艺术感染力。这是电脑图纸无法完成的阶段性任务，也是对设计起决定性作用的一步。

我们不妨回溯一下设计手绘图的基本目标——工作交流、解释方案、抓取灵感。我的观点是，草图要快中求好，而非反之。既然你能在10分钟之内，用寥寥数笔就表达出场地的主要特征和主要的设计意图，那有什么理由为此花上10个小时甚至一星期时间去做完美的渲染，以至于当你终于完成那些"完美"的渲染之后，设计的激情、灵感就消失殆尽，这种繁复的长时间渲染在今天完全可以交给电脑来完成。

当前，景观建筑手绘教育和训练受到了前所未有的重视，各种论坛和全国性的竞图活动也相当频繁，但就我的观察，此类活动仍存在一些认识上的偏差，包括对作品的评价更多的还是关注于图面的最终效果，对于设计阐述以及快速高效的表达设计意图等方面的内容关注明显不足，大赛的获奖者和参与者也以工艺美术以及环艺专业的学生、教师为主，高水平、大尺度的城市设计及景观综合表达方面的内容涉及尤其少，这与国外许多一流景观设计机构的情况形成了极大反差，所以本书在篇首就开宗明义用不少的笔墨探讨了手绘效果图在当前景观设计中的必要性，以及更有效地利用手绘技

We might as well take a review of the basic aims of hand drawing in design: to communicate, explain, and capture inspirations. I think good is the sub-pursue on the basis of speed, but not the contrary.. Since you can express out the major characteristics and design intention with a few strokes in 10 minutes, why be bothered to spend over 10 hours even a week for a perfect rendering? Very likely that your passion as well as inspiration would be exhausted after you finally finish those perfect renderings, so do leave the trivial and tedious rendering tasks to the computer.

Now the education and training of hand drawing for landscape architecture has drawn more attention than anytime before, and various seminars and nationwide competitions are thriving in the field. However, from my observation, this kind of activities still has derivation in approaching: the evaluation of works still focuses on the final visual effect, obviously insufficient of concerning on design presentation or the quick and efficient expression of design intention; the winners and participates are mainly students or teachers of art school or environmental art; the outstanding large-scale urban design and comprehensive landscape presentation are rarely seen, which indicate a huge gap between Chinese designers and many foreign first-class landscape design studios. So this book discussed the indispensable role hand drawing plays in contemporary landscape design at very beginning in more words than necessary, followed by hand drawing training skills for more efficiently serving landscape planning and design. All the illustrations in this book come from my own project designs. I try to demonstrate expressive methods and skills of quick landscape hand drawing in every detail and the whole process, wishing it would help improving the hand drawing technique of contemporary designers. Meanwhile, I believe those hand drawings which are directly drawn from real projects can certainly present

艺服务于景观规划设计的种种训练方法。并致力于用作者亲自设计完成的大量设计项目为例，从细节片段到作图全过程展示，详细介绍了快速景观手绘图的表达方法和技巧，以期对提高当代设计师的手绘技艺有所帮助。同时，这些直接来自于实际工程项目的大量手绘图纸，也向读者展示了景观手绘图在当代景观设计项目中的广阔的应用天地。

本书最后部分，将展示一种可以与现今业已普及的电脑渲染图并行的手工快速渲染技术，探讨将手绘图作为展示设计成果的手段之一的可行性，这一点，尤其在景观制图中仍将是重要的方向之一。我认为，高度成熟和个性化的手绘图在表现效率、产品艺术性等方面毫不逊色于电脑制作。在诸如意境表达、多种意象展示等方面比电脑渲图更快捷、更切题。目前国外的许多一流景观设计团队中，手工渲图仍是得到普遍认可和高度成熟的表达途径之一。我们因为认识和教育等方面的偏差，曾一度荒废的手工渲染技术，有必要得到重视和一定程度的恢复，这也是撰写本书的目的之一。

作者于两年前出版的图集在广大学生和设计师中的反响甚强，他们对图纸的印刷等方面提出了更高要求，此书出版，是对前集的补充和完善，希望能更好地服务于设计师和同学们的需求。以此代序，以飨读者。

本书在编撰过程中得到了杨琦、松华女士的热情帮助和支持，没有她们的努力，这本书是不可能问世的。

本书受：中央高校基本科研业务费专项资金资助（项目编号：TD2011-31）

王劲韬
2011 年于北京清华园

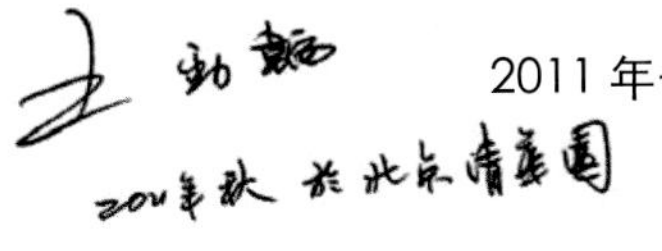

the readers a larger stage for the application of landscape hand drawing in contemporary landscape design projects.

The last part of this book tries to demonstrate a quick hand rendering technique that can be paralleled with the widespread computer rendering, and explore the possibility that hand drawing would serve as a way of design achievements presentation, which, would still be an important direction of development especially in landscape drafting. I think a highly matured and personal hand drawing is not inferior to computer drawing in expressive efficiency and product artistry, and even faster, more convenient and more accurate than computer rendering in expressing mood or certain imagery. Nowadays, hand rendering is still a universally received and highly matured way of expressing among many top foreign landscape teams. Also I hope the once abandoned hand rendering techniques shall be noted and renewed from the deviation in recognition and education, and that is another purpose of this book.

Since my book published two years ago received a complimentary feedback among students and designers, they kindly put forwards such higher demands as for the printing of drawings. This book is published as a complement and upgrade to the former. Humbly I wrote the above, as a foreword to the readers.

Thanks to Ms. Yang Qi, editor of Ifeng Press and Ms. Song Hua, thanks for your passionate help and support. Without your effort, this book may never greet the readers.

Supported by the Fundamental Research Funds for the Central University (No. TD2011-31).

The Author: Wang Jintao

目录
CONTENTS

005 序言
PREFACE

005 作为说明和交流工具的手绘图
Hand drawing as a tool of illustration and communication

008 作为设计思维过程的展现和具体化的手绘图
Hand drawing as a specified presentation of designing process

013 手绘图的未来
The future of hand drawing

020 第一部分 基础训练
CHAPTER ONE Basic Training

022 树的画法
The Depiction of Trees

046 石头的画法
The Depiction of Rocks

048 水、天空的画法
The Depiction of Water and the Sky

050 水、溪流的画法
The Depiction of Water and Streams

054 建筑及城市意象的画法
The Depiction of Architectures and Urban imagery

061 配景丨车、船小品
Ornamental Scene: Vehicles and Boats

064 人物的画法
The Depiction of Human Figures

066 第二部分 过程图解析
CHAPTER TWO   Analysis of Process Drawing

066 过程平面图
Process Plan

074 过程鸟瞰图的画法
The Drawing of Aerial View

080 过程建筑的画法
The Drawing of Architectures

085　第三部分 项目说明
CHAPTER THREE　Project Description

086	河北清河县水系规划 Planning of the Qinghe River in Qinghe County, Hebei Province	
090	唐山水处理厂景观设计 Landscape Design for the Water Treatment Plant in Tangshan	
092	黑龙江五大连池安置中心镇城镇中心博物馆广场 Central Museum Plaza in the New Settling Town of Wudalianchi, Heilongjiang Province	
094	包头幸福广场 Happiness Plaza in Baotou	
096	曹妃甸科技公园设计 Planning of Caofeidian Technology Park	
098	以色列耶路撒冷城市景观设计国际竞赛 International Urban Landscape Design Competition in Jerusalem, Israel	
106	中国园 China Park	
110	博物馆园 Museum Park	
111	河北三河市漱玉文化公园 Shuyu Cultural Park in Sanhe, Heibei Province	
114	北京台湖镇新城绿地景观规划 Landscape Design for the Green Land of the New Town of Taihu Town, Beijing	
116	迁西滦河——滦县东岸规划 Planning for the East Bank of the Luanhe River, Luan County	
119	唐山雕塑公园 Tangshan Sculpture Park	
124	秦皇岛东郊公园 Eastern Suburb Park in Qinhuangdao	
126	秦皇岛某培训中心 A Training Center in Qinhuangdao	
128	秦皇岛洋洋花海 Yangyang Flower Field in Qinghuangdao	
130	辽宁铁岭新城凤冠山景观设计 Landscape Planning for Fengguan Mountain in the New Town of Tieling, Liaoning Province	
132	唐山南湖——唐胥路 Tangxu Road, Nanhu in Tangshan	
134	唐山南湖生态城——环城水系及新开河扩湖景观 Landscape of the Urban Hydrographic Net and the Enlarged Lake in Nanhu Eco-Town, Tangshan	
138	唐山南湖生态城规划 Urban Planning for Nanhu Eco-Town in Tangshan	
141	唐山南湖生态城——青龙河景观 Landscape of Qinglonghe River, Nanhu Eco-Town in Tangshan	
143	唐山南湖生态城——市政广场 Municipal Plaza in Nanhu Eco-Town, Tangshan	
150	唐山植物园规划 Planning for the Tangshan Botanic Garden	
158	铁岭主题公园 Theme Park in Tieling	
172	潍坊白浪河 Bailang River in Weifang	
180	庭院景观设计 Landscape planning for a courtyard	
184	山西朔州七里河景观规划 Landscape planning for Qilihe River in Shanxi Province	
190	武汉东湖风景区规划 Planning for the East Lake Scenic Resort	
192	北京师范大学中心景观设计 Central Plaza for Beijing Normal University	
194	淄博农业园景观规划 Landscape planning of the agricultural garden in Zibo	
196	西班牙马德里巴德维巴斯公园 Valdebebas Park in Madrid, Spain	
198	景观手绘表现图赏析 Analysis of Hand Drawn Landscape	

第一部分

CHAPTER ONE

基础训练

Basic Training

　　本书所有内容，包括技法举例部分，全部取材于本人近几年来亲手设计的城市规划和景观设计项目，既是作品集，也是对近几年设计项目的经验总结，既为回顾项目得失，也为读者提供一个最接近于当前中国景观建筑师执业状况和要求的范本。本书内容侧重于快速、意象化景观表达，简化了一些纯绘画专业中不可缺少的基础环节，比如将物体的明暗光影的基本面由传统的高光、受光、交界、暗部、反光等五种大色调简化为单一色调，并强化物体的阴影，突出物体在空间中的落位，以便做到既快又清晰地表达空间层次。作为本人最近几年来在设计草图和意境渲染图方面的总结回顾，本书提供了一个独立于普通景观制图教育框架之外的新观点，它总结了本人在课堂教学中感受到的，学生们在能力培训方面的一些"短版"和缺项，并提出相应的训练方法，也从一个职业设计师的角度为广大学生和初涉景观设计行业的年轻景观设计师们，提供了一个适用于国内景观设计业发展要求的训练框架和较粗浅的初级范本。

All the contents including illustrations for skills come from the urban planning and landscape design programs that I made myself. This book is not only a collection but also a conclusion of my past experience in design programs of recent years, aiming to provide the readers with demonstrations as close to the real situation and demands in present China as possible. This book focuses on quick and imaging landscape presentation which simplified certain indispensable process in drawing, such as I simplified the traditional five tones of highlights, light side, shadow edge, dark area, and reflected light of the basics of bright, dark, light and shadow into a single tone, meanwhile strengthening the shadows to accentuate the space setting of the objects, thus to express the space layers quickly and clearly. As a conclusion and review of my designing sketch and imagery renderings of recent years, this book applies a new idea that is independent from ordinary landscape drafting education, summarizes the shortcomings

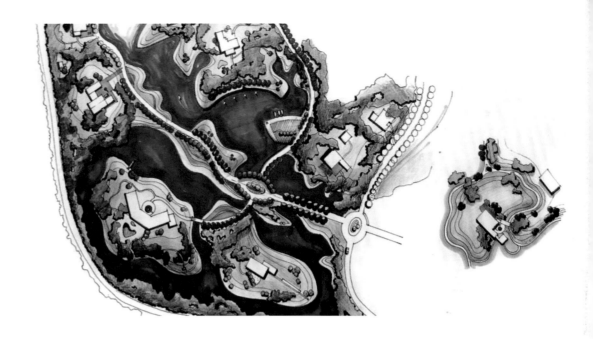

本人的理想是将设计草图之"草"、之快捷与设计所追求的达意和美联系起来。用翻译家严复的观点看,就是在实现信(真实)、达(表意)的前提下,追求一种"雅","雅",其实就是以最少的笔墨表达尽可能丰富和准确的意境,这是一种四两拨千斤的巧,更是手头、眼力和艺术素养的综合体现。

本人的意图也很明了了,让初涉此行的人在学习阶段就形成这样一个印象:设计阶段所作的图纸,目的首先是为了帮助自己和他人理解方案并便于各方交流,只是个阶段性成果而不是最终效果展示。工作草图帮助设计师说明许多设计意念方面的内容,其表现形式往往是一种特定场景以及加入了主观感受和取舍以后的清晰化的场地认知。像布朗先生所做的那样,发现土地的潜质,或曰土地之美。而这种潜质或特征,最好是在设计草图阶段就通过设计师之手快速抓取并表达出来。也正因为如此,本书中所选作品以彩铅和马克笔混合制作为多,特别强调快速、达意和交流性特征。

of students in training that I recognized in the teaching practice and relevant training methods, and also provides a training frame and preliminary demonstration in accordance to the development of domestic landscaped design for students and inexperienced young landscape architects from a professional designer's view.

Ideally, I wish to combine the convenience of quick sketch with the poetic imagery and artistry what design pursues. In translator Yan Fu's words, this is to seek "elegance" on the base of "faithfulness" (real) and "expressiveness" (representative). In my drawing understanding, "elegance" means to express the poetic imagery as abundant and accurate as possible in the least strokes, which derives from a lever-handling comprehensive skill of handcraft, insight and artistic literacy.

And my intention is very clear that each newcomer to this field may have an impression while learning that any drafts in designing, with its initial purpose is to help themselves and others to get a better understanding of and multipartite communication about the plan, is a periodical accomplishment rather than final effect presentation. As a carrier helps to illuminate the concepts in design, sketches are usually present as certain scenery with sharpened site recognition which contains subjective feeling and selection. What Joseph E. Brown did, is to discover the potential, or the beauty of land, and this potential or characteristics would best be quickly grasped and presented out by designers in the sketch process of design. I strongly recommend this kind of drawing. For this purpose, the works I selected in this book are mainly drawn with color pencils and markers, which have their natural advantages in the quick and expressive communicative drawing.

注释：
左图为常见的组群植物表达方式，多运用于中小尺度场地，总的原则是大面积和谐、点状对比。图中道路多是在完成后用涂改液加画而成。本组中，上下层植物之间的相互映衬，深浅色组群交替使用，大面积留白的草坪场地给画面带来空灵之感和必不可少的虚实对比。为避免画面过于拥塞压抑，这种疏密对比既是场地功能布局使然，也能起到调节气氛、突出重点的效果。

left: This is a common expression of groups of plants which is often used in small-scale or middle-scale site. The main principle of this kind of drawing is harmonious in large area with small punctiform contrast. Most roads in this picture were added by correction fluid after the completion of the picture. In this drawing, the upper and lower layer of plants are in the comparison with each other; dark and light colors are used in turn; a large area of blank is left in the lawn to bring a sense of emptiness. The contrast of solidity and emptiness is essential. To avoid the screen being too crowded, the contrast of the emptiness and solid is not only determined by the site layout, also has a role to adjust the atmosphere of the picture and to stress the important part.

■ 树的画法   ■ The Depiction of Trees

平面的植物

景观绘图中植物是最主要的表现内容，往往占据画面的主体位置，景观图中的前景、中景、近景几乎都是以植物表现为主。所以，景观绘图，无论是平面还是透视，均可以从植物开始，从一棵树到一组树（林），到整个园林逐步加以训练。

通常大小两种尺度和比例下的平面图表达方式是有所区别的。前者由于元素多、层次复杂，对于单个植物（或树圈）描写均以简约为佳，并需注意到树圈组群的明暗及色彩冷暖，避免平均对待而导致画面趋于涣散杂乱。

这里展示的图例以中小尺度的景观平面为主，强调植物组群的整体性和不同植物在色彩、明暗方面的变化。譬如，对于农场、酒庄一类的传统景观，在表达时，就可以加入一些主观色，如暖棕色等，以突出场地的历史感和时间上的一种延续性。而在一些带泳池或滨水的住区规划平面中，主观色完全可能是蓝灰一类的冷调子。但无论如何，各种色彩之间需有协调，各组群之间需有对比，形态大小方面需有微差及渐变，总的效果则可以

Plane Plant

As the main subject of a landscape drawing, plants often occupy the major position in a picture: from front view, mid-view, to nearby view. In this sense, either plan or perspective view, the training of landscape drawing often starts from the drawing of plants, step by step from a tree to a group of trees, and finally to the whole garden.

The ways of expressing two plans in different scales are usually distinct. Large-scale plan prefers a simple description of single plant (tree circle) for its various elements and complicated layers. Be aware of bright or dark and cold or warm in color of tree circles groups and avoid homogenized approach, or the picture would be disordered and pointless.

The following illustrations are mainly medium or small scale landscape plans, which emphasize the integrity of plant groups and the changes in colors and chiaroscuro of different plants. For example, when drawing traditional landscapes such as farm

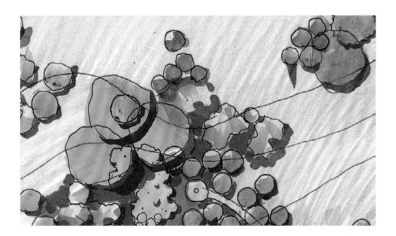

注释：

旨在表现一个带有中东文化特点的花园。特别设计的当地植物椰枣、无花果、大麦、橄榄和场地原有的柏树形成明确对比。

留白的道路和深蓝色的水渠网形成呼应，却又绝不混淆。棕黄色的基本背景色点出场地的人文特征——一种发轫于荒漠之上的发达的农业文明和以水渠为特征的农业文化。对场所的记忆和文化意义的展现均得力于恰当的色彩运用和元素提示。

The drawing intends to express a garden with Middle-eastern cultural characteristics. Local plants there such as dates palms, figs, barleys and olives were specially drawn to be compared with the original cypresses in the site.

Roads which were left as blank and the dark blue drainage network respond to each other but without any mix. The brown background of the whole picture indicates cultural characteristics of the site—a developed agriculture civilization thrived on the desert and the culture signatured by drainage system. The proper use of color and elements contribute to expressing the impression and cultural meaning of the site.

某古典别墅花园　　a classic-style villa garden

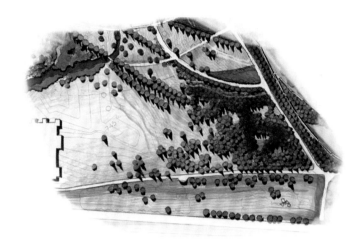

某古典别墅花园　　a classic-style villa garden

相对统一。

本节共列入 3 组中小尺度平面图和大尺度平面图，分别从某古典别墅花园，以色列国家历史花园，宾馆总统套房花园和滦河迁西段新城规划等项目总平面图中节选局部，说明了以上本人提出的色相差异与文化意境的某种关联性。读者可以在以后的景观创作中细细品读体会这种来自色彩的隐喻特征。

在具体技法层面有两点需要注意：

1. 平面图中的投影很关键，不可简省，而各种植物的平面投影也不尽相同，如阔叶树为圆形平滑投影，针叶树则为锥形投影，一张图中投影方向须保持一致。投影是使图面厚实稳固的前提，图中植物通常只有在加上投影后才会"落地"，画面才可能有一个相对统一的语言。投影用色宜采用——黑或深灰，它们在多种色彩相对比中，是极有效的调和剂。

2. 组群对比，对于1：1000以上的大尺度场

or Chateau, we can add some subjective colors like warm brown to accent its historical atmosphere and continuity in time. In other cases, the subjective colors may shift in a cold tone of blue or grey, especially when drawing waterfront residential plans or plans with swimming pools. Anyway, a good drawing calls for the coordination of colors, the contrast between groups, tiny differences and gradients in shape and size, and relative unification of general effect.

This section contains 3 groups of mid-small scale plans and a group of large scale plans. In the general plans of a classic-style villa garden, Israel national history garden, a presidential suite garden, and new town of Qianxi County in Luanhe River section, I selected some parts to illustrate the certain relevance of hue differences and cultural context that I mentioned before. You may carefully read and feel the metaphor of colors in future landscape creation.

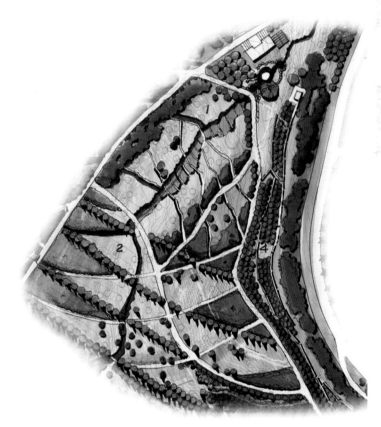

某古典别墅花园　　a classic-style villa garden

以色列国家历史花园　Israel national history garden

地规划，通常以云线片林和点状列植的对比，体现不同组群和设计中的地位。比如，你可以用云线表现背景林、线状林，而用更具体的树圈表现重点设计的树阵和点景大树等。背景若使用冷绿色，重点描绘的点植部分则可以用稍暖色调，色彩也更饱和，以拉开组群的层次，又不至于过于突出。

小尺度场地上，平面植物的对比手法则更为丰富，通常大乔、灌木及地被可多层叠压，最上层的大树仅以单线和阴影处之，空白处以树下灌木和地被层层掩映，结构清晰，层次丰富。第三组平面中的一系列日式花园表达中，就采用了这种多层叠压的方式，深色的地被层为所有的上层植物、置石、水景提供了统一背景，使画面元素丰富而不零乱。

1:500以下的小比例、小尺度空间在植物表达的形式方面比大尺度景观更为丰富。一般要求有明确的轮廓和栽植定位点。对于小于 1:100 的庭院空间手绘图，还需要表现出分枝、分叶等变化，用以明确区分阔叶、针叶树种，树下的地被花草也多要以带定位点的圆加以表达。这部分内容在现在可由 CAD 自动生成，大型图库中的各类植物图块可供选择搭配，手工只需稍做后期加工，一般都能取得较好的效果。

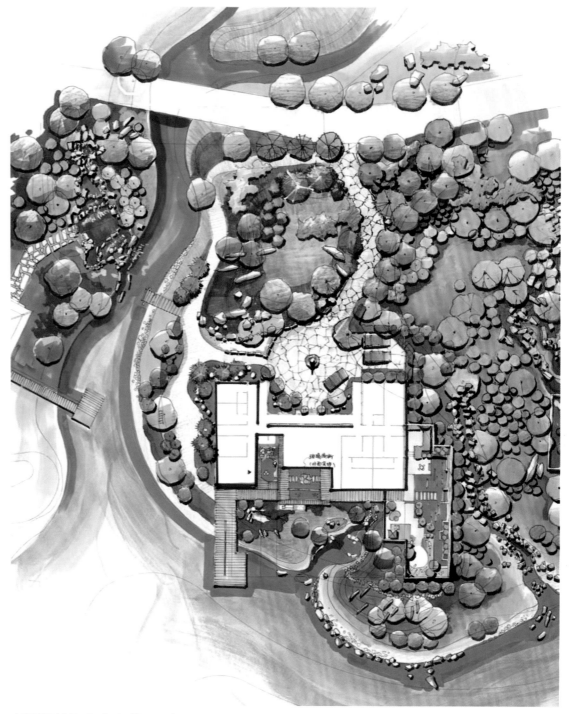

以色列国家历史花园　Israel national history garden

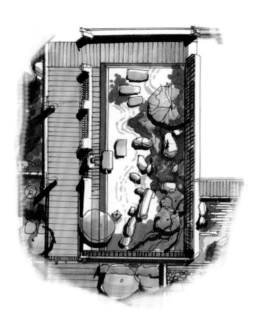

宾馆总统套房花园　　a presidential suite garden

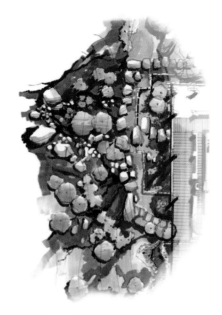

宾馆总统套房花园　　a presidential suite garden

宾馆总统套房花园　　a presidential suite garden

注释：
　　上图为典型的小庭院平面表达。平面线稿底图较为复杂，仅景石一项就有规则的敷石、半规则的抛石、汀步和完全自然化的石滩、石组等，植物则包括乔、灌、地被层层覆压，几乎密不透风。所有的秩序、条理均是由后期色彩所规范，最强烈的对比由日本庭园的两大特色元素——白砂和苔藓地衣形成，次级对比为槭树、木夹板等中间层次，大量的日本松全由深色地被衬托，层层叠加、多而不乱。

above:This shows a typical expression of small courtyard layout. The original line drawing plan is quite complex. Only for the types of landscape stones, there are regular deposited stones, semi-regular ripraps, stone pathway through water and fully naturalized rocky beach and rock groups, etc. For plants, there are different kinds of trees, like arbors, shrubs and groundcover plants overlapping layer by layer, almost airtight in the drawing. All the orders and rhythms are given by colors afterwards. The strongest contrast in color is resigned to the two hallmarks of the Japanese gardening elements—white sand against mossy lichens. The secondary contrast is among maples, plywood and other intermediate level. A large number of Japanese pines are set off by dark groundcover plants, overlapping without being chaotic.

Note two points of specific skills:

1. Cast shadow is so important in plans that it can never be simplified or omitted. Since different plants have different planar projection, e.g. broad-leaved trees have round smooth shadows while needle-leaved trees have tapered shadows, and the direction of cast shadow should be unified in the same picture. Projection lays the foundation of a stable and solid sense of a picture. Plants would look like dangling in the air unless they are added cast shadows, and the whole picture being unified thereupon. The color of shadows should be of uniform neutral color (black or dark gray), for it blends effectively with various contrast colors.

2. The contrast of groups. For large-scale site planning of over 1:1000, we can use the contrast between cloud line woodlots and punctiform planting to accent the distinct positions of different groups, e.g. background trees and linear trees can be presented by cloud lines and accented tree groups and big trees can be showed by more specific tree circles. If we choose cold green as the background color, the accented spot trees shall use some warmer and more saturate colors to separate but not protrude different group layers.

In small scale site, the contrast skills for plane plants are even more various. The multilayer of large trees, shrubs and ground cover plants are overlying each other with the top-layered trees presented by single line and shadow, and the blank space filled with shrubs and ground cover plants under the trees, creating a clear structure but abundant layers. In the third group of plans, the series of Japanese garden plans adopt the multilayer overlying skill to make the picture abundant and but not messy: the deep-colored ground cover layer provides a uniform background for the upper plants, stone layouts and waterscape.

The plants in small scale space of under 1:500 are more abundant than large scale landscape in respect to way of expression. Small scale landscape plan usually requires clear outlines and planting anchor points. For courtyard space sketch under 1:100, the branches and leaves should also be specified to separate the broad-leaved plants and needle-leaved plants, and the ground cover plants are usually expressed

滦河迁西段新城规划　　new town of Qianxi County in Luanhe River section

注释：
　　大尺度植物平面，重在表现绿带、水网与城市建设区之间的穿插，建筑留白、铺装的各种灰色填充，突出深色林带与浅色开阔草地（市民活动场地）之间的对比。

　　这类平面中，树圈数量极大，宜成片描绘，需要重点突出的部分，少量单点。超过 1:1000 以上的大比例图纸中，平面植物不要分出树圈明暗，单色平涂即可，否则易于混乱。同类树形可连片加设投影，形成明确的栽植意象。总之，由于平面大，元素多，亮点也多，各种树木描绘均宜平整，不宜用色对比过强，以致整个图面处处"闪光"，其实是一无是处。

This is a large-scale plan with plants. The stressed point is to express the intersection between the green corridor, water network and the urban construction area, extruding the blankness left for buildings, various grey fills of the pavement, and the contrast between the dark forests and the light-colored parkland which is a space for public activities.

There are a great number of tree circles in this type of plan which should be drawn in groups and the few necessary parts are emphasized respectively. When the scale of the plan is bigger than 1:1000, trees should be painted in monochrome and the dark and light parts of the tree circles should not be drawn distinctively to avoid chaos in the whole picture. The shadows of the trees of the same form can be drawn together to form a clear plants imagery. In brief, in a big site with a big plane, lots of elements and highlights, all sorts of trees should be drawn neatly and homogeneously without strong contrast in color. Otherwise, the whole picture would be flashing everywhere, thus ends in nothing.

　　主干大树、点景树由于体量较大，应视为球形处理，方式类似于素描排线画法，在精细描绘中，除了以上所说的分层叠加，投影落地，以及色彩、层次等方面，对于较密植的平面，还应该注意树冠之间的相互避让，分枝重叠处要特别注意前后层次，一般树圈交叉的情况下，只需表达其中之一即可，不要过于重叠以至于使画面层次杂乱。

　　大型阔叶树也可以只表现大体树形轮廓和投影，重点表现树下空间。

　　小比例庭院空间植物色彩处理尤为重要，基本经验是：大树用色宜平和，树下的小型植物宜鲜艳突出，最后阶段，可以单色阴影归拢多个层次，即可取得和谐及对比的效果。这类小比例植物平面表现图的色彩对比，在景观师布雷马科斯的作品中表现最为突出，显示出他早年从事印刷设计工作的一些色彩方面的感悟，同时，他的配色模式与他所表达的热带庭院植物主题也甚为契合，可资参考。

through circles with anchor points. Of course now these parts can be automatically generated by CAD, and various plants blocks are free for your choice in large picture data, with some of your modification, you can always achieve better effects.

Major trees or accent trees are often too big that they should be treated as balls and in a way similar to hatching in sketch. In specific drawing, for intensely-planted plan, besides being aware of multilayer overlying, projection grounding, colors and layers, you should also notice the avoidance among tree crowns and the order of overlying branches. If tree circles intersect each other, carefully express out one of these is enough. Don't let too much overlying disturb the ordered layers of the picture.

For large broad-leaved trees, expressing out the general outline and projection would be enough, since the point is the space below the tree.

The coloring of plants in small scale courtyard is particularly important. The basic experience is, using moderate colors for big trees and vivid colors for small plants under the tree, and gathering the layers with single-colored shadow to create a contrastive but harmonious effect. Landscape Architect Burle Marx's works give a good example of the color contrasts of plants in small scale plan presentation. Containing his own understanding and perception of the printing design experience of his early years, Marx's works are worthy of reference for his color-matching mode corresponds so well with the theme of tropic courtyard plants.

注释：
本组图，上下层植物之间的相互映衬，深浅色组群交替使用，大面积留白的草坪场地给画面带来空灵和必不可少的虚实对比。避免画面过于拥塞压抑，这种疏密对比既是场地功能布局使然，也能起到调节气氛、突出重点的效果。

This is a common expression of groups of plants which are often used in small-scale or middle-scale site. The main principle of this kind of drawing is harmonious in large area with small highlight with contrast. Most roads in this picture were added by correction fluid after the completion of the picture. In this drawing, the upper and lower layer of plants are in the comparision with each other; dark and light colours are used in turn; leave a large area of blank in the lawn to bring a sense of emptiness. The contrast of solid and empty is essential.

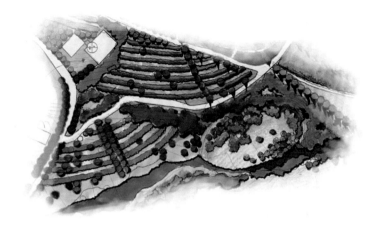

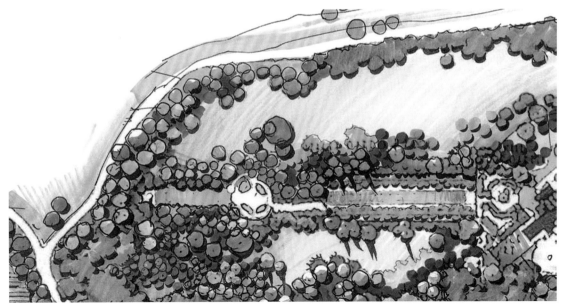

透视植物

本节简单解析了人视点透视图中的植物表达和大尺度鸟瞰中较为概括的树群、草坪等元素的表达形式。总的观点是，我们在绘画和设计上的许多构图原则，如不规则、非对称、均衡特性等，均可以在一棵树的表达上充分体现出来。就具体表现对象而言，单株的点景树，组群式的树林、树阵，以及远景成片的背景林等，在表现的虚实与手法应用上往往是大相径庭的。本节按点、线、面的顺序一一加以论述和举例

Plane Plant

This section gives a simple analysis of the way of expressing plants from human's viewpoint in perspective drawings and more general elements, such as trees and lawns, in large-scale aerial view. Generally speaking, many principles on drawing and designing, such as irregularity, asymmetry, and balance, can be fully expressed out on a single tree. For specific object, the operating techniques and ways of expressing emptiness and solidness of a single accent tree, tree groups, and background forests in distant view, are usually very distinct from each other. In this section, I will discuss and demonstrate one by one from point, line to plane.

鸟瞰图中的植物表达

这类图纸通常规模变化较大,从几公顷到几十、几百公顷,甚至整个城市片区的鸟瞰都要求在一张图纸上完整地表达出来,这往往对中远景的表现技法,以及表现的深度会提出很高的要求,同一植物在不同距离和视点高度上,也会表现出几乎完全不同的形态和色彩。正确的透视变形、大小变化和色彩冷暖以及饱和度变化都会有助于更好地体现出场景的远近层次和变化。

一般而言,低视点、中等尺度的公园可以表达出树木的品种和组群的基本形态。同时,组群色彩上也可以稍作对比变化。

随着场景范围增大、视点抬高、表现的焦点逐步转移到整体形态,用色也需逐步统一,将变

Expression of Plants in Aerial View

The scales of the objects in aerial view are various from hectares to hundreds of hectares, and even a whole urban area. This usually requires for professional skills of expressing the depth of mid-distant view. One should aware that the same plant presents totally distinctive shape and color in different distance and viewpoint. The proper perspective distortion, size transformation and color saturation changing from cold to warm can help a better presentation of different layers and depth of scenery.

Generally speaking, the lower view mid-scale parks usually bring out a clear expression of the species and basic forms of the tree groups. Meanwhile, a little color contrast or change between different groups would be better.

Along with the increasing scale and rising view, the concentration of a presentation shift to the overall

化和对比留给更大尺度的组群，如背景与前景的渐变，水与植物的对比，天空与树群的对比等。

正如中国传统绘画中讲的"远树无根"，这类片林树群，可以以树冠和树球的体积感表达为中心，鸟瞰的树干不仅大大缩短，而且几乎与投影混成一体，草图中只需使用排线平铺，表达出林带的体积感即可。除了近景以外，树群一般不需多加刻画。

大尺度鸟瞰由于涉及的元素较多，空间极大，许多绿块、树群、大片林都有可能缩减为画面上的一个点，因此，树群之间，树与草地、铺装等元素的统一，往往要依靠一种整齐的排线加以归拢，这种画法类似于光影素描，统一和谐，以空间层次的正确表现压倒一切，与层次和整体统一相抵触的任何细节，无论画得多么精彩，都有必要用排线"压"下去。这就和当年罗丹砍掉巴尔扎克雕塑上那只精美绝伦的大手是一样的道理。这一点初学者往往不易做到，但确确实实是做好大尺度鸟瞰表达中最重要的技法和极好的创作习惯。大鸟瞰对植物表达的总体要求是统一整体，服从于空间表现的需要，树木本身的特点稍作提示即可，不宜"抠"得过死。投影宜统一布置，近景深、远景浅，以表达出空间距离和"云气"等场景特征。

appearance, the colors become more unified, leaving the changing and contrast to groups on larger scale, such as the gradient transformation of background and front view, the contrast between water and the plants, as well as the sky and the tree groups.

"Trunks of distant trees are invisible". Same as what in traditional Chinese painting, the distant tree groups are expressed by emphasize the volume of treetops or tree balls. The shortened trunks are almost blending with the projection in aerial view, and you just have to use hatching to spread out the volume of tree belts in sketch. Tree groups are usually free from exquisite drafting except in nearby view.

For the various elements and vast space of large scale aerial view, many green blocks, tree groups and woodlots may be contracted to a single point in the whole picture. Therefore, the unification between tree groups as well as trees, grassland and the pavement can be assembled by a regular hatching, which is similar to light and shadow sketch. We can't emphasize too much the correct expression of unification, harmony and space layers of a picture; any details that contradict with the wholeness and layers should be unified by hatching regardless how wonderfully it had presented. This is for the same reason as Augeuste Rodin cut off the exquisite arm of the sculpture Honore de Balzac. Though maybe difficult for beginners, is truly the most important technique in drawing large scale aerial views and also an extremely good creating habit. Being unified is the general demand for plants in large scale aerial views, thus to serve the requirement of space expression and slightly indicate the features of plants would be enough. As to the projection, to express the scenic characteristic of distance or "air" etc, it would be better to planning on the whole with dark shadow for nearby view and light shadow for distant view.

注释：
本组图为低视点鸟瞰的典型样式，也有人称之为平行鸟瞰，形式类似于人视图中的一点透视，构图较稳重，均衡，宜画得确切扎实，也要重点分明，避免涣散。

注意事项：
格局：此类小鸟瞰景观一般可表现到场地环节，图面需反映种植设计所规范出的各种场地的空间特色，诸如开放的草坡、封闭的林窗等场地形式，可以在小鸟瞰中一并表达出来，做到用艺术化的手法表现出系统的场地功能配置概况。

本组图例反映的是唐山某中央公园三块具有代表性的绿地，分别作为纪念性雕塑展场，一般市民游园场所，以及滨水综合绿地（兼具运动场、休闲绿地、防护绿地等多种功能）。树群组团的体量、色彩搭配均迎合了未来场地使用上的需求，如雕塑展场用姿态松等常绿树背景与樱花林映衬，形成多组纪念性的环境背景，以适应雕塑艺术展示的需求；右下的临水场地，则以大片林带半围合空间，形成明确的方向性空间，向大湖一侧开放，这些空间的设计意图，均可以通过明暗及光影处理予以强化。

点景：即便是较大尺度的鸟瞰图，人仍然是重要的配景元素。本人习惯于在所有层次完成后，直接用白色涂改液"点"出人物。此外，大面积的阔叶林中以重彩点出针叶树，使之高出树群，往往能起到点睛之笔的效果。

The three sketches of this group present the typical pattern of bird's view on a low viewpoint, which is also known as parallel bird's-eye view. It is similar to the one-point perspective in human's view. The composition of this type of perspective is with more stability and more balance. The brushwork should be exact and solid while the focal points should be stressed and free from scattering.

Notice:
This kind of small-scale aerial view can usually depict the site at the level of details. The drawing should present the spatial characteristics of various sites regulated by planting design, such as open turf slopes, closed canopy gaps and other elements. So that the site's functional configuration can be systematically expressed in an artistic way.

The groups of legend depict three typical green lands in Tangshan Central Park respectively, which are monumental sculpture exhibition site, the general public leisure garden and integrated waterfront grass land (also serves as stadium, leisure space, green buffer, etc.). The volume and color of tree groups are in accordance with the requirements of site for future use. For example, sculpture exhibition area is decorated with pines with special poses and other evergreen trees silhouetted against cherry woods to render the monumental environmental background and to meet the needs of art exhibition. The waterfront zone on the bottom right of the picture is semi-enclosed with forest belts to create a clear guiding sense to the lake. These design intentions can be enhanced by the effect of shading and light and shadow.

Scene decoration: even in a larger-scale aerial view, people are still important decorating elements. I'm used to directly "pointing" out the people with white correction fluid out after the completion of all layers. In addition, you can put another layer of thick color to stress conifers and to make these them above the large area of broad-leaved forest. This effect can often play a role as a finishing and highlighting touch.

注释：此组图为南湖总体鸟瞰图节选，为近景岛屿和树群的概念性表达，图中除了一级干道以外，其余线性空间均予以省略，以服从植物为主体的表现图纸的需要，其他元素也尽量精简，以保持画面的整体性和纯净的效果，较大的公共建筑和景观构筑物直接空白，也可适当用浅色马克或墨线排线，以协调画面的整体感觉。

These two pictures are selection of the overall aerial view of South Lake, which is the conceptual expression of islands and tree clusters in a nearby view. Except for the first-grade main roads, all the other linear spaces are omitted to emphasize plantings as the main theme of the picture. And other elements are painted as concise as possible, in order to maintain the integrity and purity of the picture. Larger public buildings and landscape structures are directly left blank (we may as well use light-colored marker or ink hatching) to coordinate the general atmosphere of the image.

注释：
　　休闲农庄的树林用色宜偏暖褐色。图中的西式酒庄附有大面积的葡萄种植园和周边防护片林，两类林一为规则、矮小，一为高大、自然种植，在色彩深浅和色相等方面都宜做出明确区分、不要混同，远处的背景林带要渐次虚化，以突出场景的深远感和层次感。

The warm brown color seems to suit the woods of a leisure farm. In this figure, the western style wine chateau has a large area of grape plantation and the surrounding shelter forest. Of these two kinds of planting, one is regular and short and the other is tall and naturally grown. The color depth and hue of the two types are advised to make a clear distinction without mix or confusion. The distant background forest should blur gradually to create the sense of depth of the screen.

注释：
　　树林与城市背景搭配的鸟瞰要先明确主次。若以表现绿地公园为主，则城市背景色彩要低调并统一，可以只用单色(单一灰色系马克最佳)分出亮暗面即可，其余细节全部省略。(若以表现城市建筑组群为主，则要有意将前景林色度和明度降低，用统一的前景衬托丰富的城市意象。) 统一画面的一个简单易行的方法就是在马克笔着色完成后统一加上墨色排线。排线完成后，若觉得过于沉闷，还可以用涂改液加画点景，以"提醒"画面，这和大面积田野或草地上加画点景树道理是一样的。

The main theme has to be clear first in a bird's-eye view with city scene mixed with forest. If the theme is green parks, all the urban background colors should be understated and unified. Only dark and bright sides are distinguished with monochrome (mono-gray marker would be the best choice) while all the other details are omitted. Gray colors in the same serious may be the best choice. If the main character is the building groups, the chroma and lightness of the forest in front view should be lowered intentionally. A rich city image will be set off by a unified close view. A simple method to unify the picture is to add ink hatching onto marker colors. If it turns out to be too dull, the correct fluid can be used to point out some spots as decoration to "remind" the picture. It is just like adding some decorating trees in a large area of meadow or field.

注释：

植物布局类似大地艺术作品，更适合于用鸟瞰图表达，大面积的花田、花带色调与场景氛围要一致，由于面积极大，这些前景花田在相当程度上决定了画面的总体色调，因此用色要平、求整体效果，小块植物表现则要求突出对比。同样，色相上也需注意大面积和谐（同类色），小面积对比（补色）。

本组采用了浅黄和深紫形成极为鲜明的色彩对比，此外浅紫、粉红、玫瑰均可以作为此类"花海"主题的理想配色方案。但作为背景的树林草坡用色则宜沉着、厚实，即要有大面积重色"压"住这种强对比，形成鲜而不俗的画面效果，这种技法与植物设计的意图（常绿树搭配色叶树）也十分吻合。

This kind of plantation layout is similar to the earth's art work, which is suitable to be put in bird's-eye view. The color of the large flower fields and flower belts should accord with the atmosphere of whole scene. Due to the giant area, the front view flower fields decide the general tone of the whole picture at a considerable degree.

So it is better to color it balanced and unified. Contrast should be protruded when drawing small parts of the plantation. The same rules also fit in the hue—harmony in large area (similar color) and contrast in small parts(complementary color).

This group uses light yellow and deep purple to create a dramatic and vibrant contrast in color. Purple, pink and rose can also be the ideal coloring set for this kind of flower fields. But the forest and meadow as a background should be painted dark, thick and solid to "overwhelm" this strong contrast in the picture and to create an effect of fresh but not tacky. This technique in drawing is consistent with the design intent of the planting design (even green trees matching non-green-leaved trees).

点景大树

以下介绍人视点树的画法,分别称为点景大树、前景植物和特殊植物,这种称谓是为了表意的清晰和细节关注点上的区分。

基本内容是一致的,就是讲的树的详细画法,这也是效果图中最重要的部分。大体包含对树的认知方法、学习方法、表达方法三个步骤。

在阐述景观树木画法之前,需提醒本人一点,在树木表现这一环节,有许多细节技法可以充分借鉴水彩、素描等相邻画种的一些成熟经验,比如历史上最有名的森林画家、俄罗斯人希斯金的许多素描写生,在单株树木的刻画、树林的层次、总体氛围渲染以及大树尤其是松柏一类,与溪流、巨石搭配,构成的场景组合等方面,可以为初学者提供最好的范本。这里稍作提示,供有兴趣的读者去查阅、借鉴。

Accent Tree

Thereafter introduce the skills of drafting trees on human's viewpoint. I called them "accent tree", "front view plants" and "special plants" respectively for the purpose of separating the concentrating details and expressing more clearly.

The basic contents are still about the details in drafting, which is also the most important part in design sketch. Generally speaking, it contains three steps: recognition, learning, and expression of trees.

Before I state the skills of drawing landscape trees, I wish to notify that many details skills can be borrowed from the similar but more mature experience in gouache and sketch when expressing trees. For example, the sketches of Russian painter Ivan Ivanovich Shishkin, the most famous forest painter in history, provide with beginners almost the best demonstrations for learning the portray of a single tree, the layers of forest, the overall atmosphere rendering and the complex of scenery such as big trees (particular pine and cypress) matching the streams and large stones. May the interested readers check and learn from them.

注释:
　　一座农业示范园的酒庄景观。重点突出前景的一棵参天大树。此树描绘细致,基本体现了上文所说的分枝、穿插、阴影及树干机理几个方面的表达要求。中景的树林就是按"冠、干和阴影"三部分来归纳、表达的,远景树林则近乎平涂,场景的深远感也由此体现出来。

This is the landscape of a demonstration agricultural chateau. A towering tree in nearby view is depicted meticulously. It shows the required skills to expressing branches, crossing, shadows and the texture of the trunk mentioned above. The woods in mid-view are conculed and expressed in the three parts of tree- crown, trunk and shadow. Forest in distant view is almost painted flat. The depth of scene is thus expressed out.

注释：
中景的树林，更带有装饰树的特征，重视树形、边缘的刻画。树林成为远山和近景的溪流、广场之间的屏障和调和色块。上图是青岛附近的一个保护区设想图，为了体现当地独有的石山、裸岩和岩生植物的独特风景，故并未按照一般风景画原则，将远景画虚、色彩变灰，而是反其道而行，但这种纯色的冲击并未使画面混乱而不可收拾，原因就在于中景的这层树林，用墨线加以统一（近于概念性的装饰树画法），用色统一（又深又沉）是以形成屏障，"压"住远景的强对比，反而产生了奇特又十分和谐的效果。

The trees in mid-view are featured with ornamental plantings which emphasize the depiction of tree's clear form and fringe. They become the screen and blending color lump between the distant mountain and nearby brooks and plaza. The picture above is an imagination concept for a protection zone near Qingdao. To express the unique local rock mountains, bare rocsk and rock plants, the principles on common landscape drawing, which require blur in distant view and gray in coloring, is abandoned. Instead, the drawing approach in this picture is quite the opposite. But the strong contrast from high purity in colors does not bring out chaos and mass. The woods in mid-view contribute to that. They are unified by hatching (similar to the drawing approach of conceptual decorating trees) and similar dark color so as to form a screen. Therefore, the obvious contrast in distant view has been "overwhelmed" A peculiar but harmonious effect is produced by that.

这里需要指出的是：功夫在画外，许多高手快速表达时，似乎不经意的涂抹或寥寥数笔的勾勒就能将一片树林表现得韵味十足，而且着墨不多、恰如其分。简单评价，觉得是一种"手头功夫"，但实际上所有你所见诸于画面的都不叫功夫，功夫在画外，你若只从画面表达上亦步亦趋去"仿"，那是学不来的。本人的建议很简单，先暂时放下你要画的那棵"景观树"，花点时间把希斯金、霍默、门采尔等人的素描仔细揣摩一番，哪怕学个半吊子，回过头来再画你的"图上之树"，你会猛然发现原来景观表现要求的树太容易做到了。有了这样的练习过程，你离"怎么画、怎么有"的状态就不远了。画树很难，所以本人要赘述这一段。切记：功夫在画外，多向历史上的名画，尤其是画家们搜集素材的素描速写稿汲取一点营养，这是成功的捷径。

Still, art of drawing comes from the outside. When the experienced illustrators make quick sketch, they perfectly express the charming of the trees but by some few random and casual like strokes, proper and efficient. Judging by the appearance it seems a handcraft, however, the really craft hides so deep behind the surface that simply simulating the outlook appearance may ends in fruitless. My suggestions is, forgetting about the object landscape tree for a second and spend some time reading carefully the sketches of Ivan I. Shishkin, Winslow Homer, Adolpyh Von Menzel etc. No matter you had understood or not, when you approach the drawing of "trees on picture", you may suddenly realize the trees are so easy to express in accordance with the demands of landscape drawing. Only through this kind of exercise can you be close to the stage of "free to draw". Since it is very difficult to draw trees, I'm going to repeat again these words: art of drawing comes from the outside. Remember a shortcut to successful drawing is to learn from famous paintings in history, especially what great painters sketched for collecting source materials.

注释：
　　就如我们在北方地区进行种植设计时遵循的一条普遍原则，为了保持冬季景观效果，需要保证一定比例的常绿树，景观图中的植物配比也符合这个原则。通常大片色叶林或落叶树林背后安排一至两层常绿树，会取得非常好的层次和色彩搭配效果。这时常绿树，如松、柏，需要有意加深，并使形式简化，形成一条清晰而又色彩沉着的背景林带，如此，即使你的画面前景再复杂，再鲜艳都不至造成零乱或艳俗的效果。松柏之下设花带，可以展现映衬之美，色叶树下再摆花丛，那就无异于添乱，这和种植设计的规则的是完全一样的。

Similar to a general rule we used in planting design in the north, a certain proportion of evergreen trees should be ensured to maintain the winter landscape effect. The painting of a landscape plan also fits it. Usually, one or two layers of evergreen trees are arranged behind a colorful forest or deciduous forest. It turns out to have a great color mix effect and the sense of overlapping. In this case, the evergreen trees like pine or cypress require darkened colors and simplified shape on purpose to create a clear but steady-colored belt of background. With it, no matter how complex or colorful the close vision is, the picture would not fall into tacky or massy. It has a good effect of setting off to put flowers under evergreen trees. But it would be redundant in the reverse case to add flowers between colorful trees. This rule is also the same with planting design.

注释：

左图是为某自然风景名胜区的保护规划所做的意向图，很好地说明了天空、水面、溪流、山崖与树木之间的映衬关系。植物的主角——松树，采用了近乎单色的简单画法，但树下的植物层次丰富，野花地被争奇斗艳，全凭大面积墨绿和黑色的松林压住色彩。

仅从画面效果看，图中的溪流、花丛、水石的表现，近乎纯粹的风景写生画，但其实这幅图所表现的中心内容是山谷溪流的自然环境给人的总体印象，这种印象或"意象"是我们表达的核心，画面元素的取舍、强弱对比均是以此为标准组织进画面的，这与纯粹的风景画是完全不同的。这就有如我们看萨金特笔下的威尼斯风俗画，更像是技艺高超的风景画，而看霍默或怀斯的风景画，则更像是赋予了某种哲学理念的"意象"图一样。这方面的鉴赏以及融汇应用的能力是需要着意加以培养的。

The figure above is the conceptual planning image for the protection program of a famous natural scenic area. It explains the setting-off relationship among sky, water, streams, cliff and the trees. The main character in planting—pine is painted nearly monochrome simply. But there are abundant of plantings under the tree and wild flowers thriving. Large areas of dark green and the black pine forest are essential to control the picture.

Viewed only at the aspect of image effect, this picture depicted streams, flowers, water and rocks in a way of pure scenic painting. But the theme of it is to deliver the overall impression of the natural environment formed by valleys and streams. This kind of impression or "imagery" is the core of our presentation so that all the elements and contrasts are selected and stressed according to this standard. It is different from a pure scenic painting. The difference between them is like the impression that we view the Venetian genre paintings by John Singer Sargent is more of the scenic paintings, or we view the skillfully created scenic paintings by Winslow Homer or Andrew Wyeth which is full of some kind of philosophy connotations. The ability to appreciate and master these art skills calls for special care and efforts.

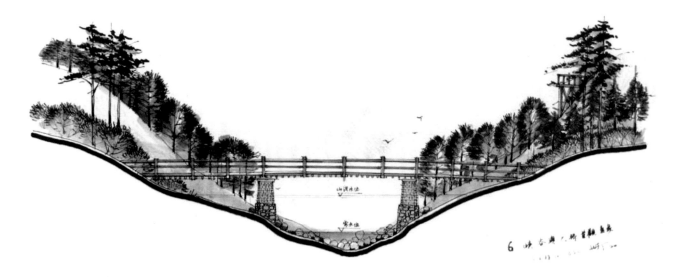

注释：
　　右图很好地说明了植物意象对空间的影响。两张图都是半小时内完成的意象小稿，除了近景的一两棵大树稍做刻画以外，中远景的树木几乎都是成团、成簇的"印象"，把所有的精彩留给了树木所围合的空间。这种小速写稿宜多练习，对于空间想象力，以及方案到透视的转化能力的增长都很有帮助。练习时，画面宜小巧，普通 A4 纸大小即可，本图例的原大仅为 A4 纸的一半。

These two pictures illustrate the influence of plant figure has on the space feature. Both the pictures were finished within half an hour as small drafts. Only one or two trees in the nearby view were depicted slightly. The other trees in the medium and distant view appear in groups or clusters. The space enclosed by trees is thus emphasized. It is useful to practice a lot with this kind of draft to develop the space imagination and the ability to transform a layout design into perspective. When practicing, the smaller screen is fine, usually A4 size. The original size of these pictures is only half of an A4-size paper.

注释：
本组图例分别展示了线和面两种近景树的表达方式，上图是以面为主的画法，宜虚笔为主，故多用彩铅画主体，马克归拢，线画为主则宜笔者为主，用线宜活、运笔要快，多用排线，使色彩对比强烈，但不致艳俗。

This group of six legends illustrates the expression of linear and plane tree in close view. The upper image shows trees in plane which employs blur lines. So the main subject is painted with color pencil and regulated by marker. The linear drawing method favors clear brushwork in which the hatches should be flexible and quick while colors should be in contrast but not flashy.

前景植物

认识树木，学习树木生长的基本特点

从表现的角度看，自然树木大体可以被分为装饰树和写实树两类，事实上在建筑和景观表现图中，这两类画法可能会被交替使用，比如近景会采用较为逼真的手法画出更写实的自然之树，而远景则有可能用较为概念的树圈或具装饰性的分枝排列加以概念性表达。

表现难度上，装饰树大于写实树。所谓装饰树，即三笔两笔表达出树种及特征的"写意"画法。画好这样的树，对于认识树木的出枝规律、总体形象往往有更高的要求，通常是在归纳和简化的基础上才可能做出正确的简笔表达。所以在学习阶段，就需要对分枝型、分叶型、轮廓型（乔木、花灌木等团状植物）以及它们之间成组、成丛的组合规律加以琢磨。

Plane Plant

Understanding Plants: Learning the Fundamental Features of Plants' Growing

On the angle of expression, the natural plants can be divided into ornamental trees and realistic trees. In architectural or landscape presentation, these two expressive techniques are usually used alternatively, e.g. nearby view would use more realistic techniques to express natural trees in reality, and distant view would be more generally or conceptually expressed by trees circles or decorative branches arrangement.

On the difficulty of expression, ornamental trees are more difficult to present than realistic trees. The so called ornamental trees, whose spices and features are often expressed by two or three strokes in a poetic way, require deeper understanding on the rules of branching and overall images of them. Its correct expression is usually on the basis of generalization and simplification. So pay attention to the rules of drawing plants with distinct characteristic on leaves, branches, and outlines (especially the cloud-shaped arbors and shrubs) and carefully study their grouping rules in learning.

特殊植物的表现

就单株树表达而言，旨在分枝，要使分枝自然合理，首先忌"平"，要在二维图中画出三维（360°出枝）的特征。对于上图那样的特大点景树而言，所有的穿插、俯仰、大小出枝，及正面枝条透视压缩、背面树条的掩映及色彩处理上，从前到后，从深灰到浅灰，使前枝突出而后枝退远，这些方法、意识都得有，此即一般画树教材中所述："光影变化、透视变化"，这八个字，剖解下来，至少应包括本人上述的这些方方面面。实际作图中，可能不需要兼顾这么多内容，即便是上图那种大树，也只需解决好出枝前后层次及主干肌理的表达，也就足够了。但在认识树木、观察树木时要有这些意识。

还是那句话，功夫在画外，景观师几乎每天都要画树，在学习阶段多花些时间揣摩这些问题还是很有必要的。

Plane Plant

As to express a single tree, the key is to make its branches as natural as possible. The first thing is to avoid being too "plane-like" to present a 3D effect (branching on 360°) in 2D drawing. In drawing particular huge accent tree such as in Fig. 1, pay attention to the crossing, pitching, branching as well as the perspective contract of the front branches, the hiding of covered back branches and their colors, from dark grey to light gray, accentuating the front branches while fading the back branches. (What I have set forth at least contained the general concepts which ordinary textbooks on drawing had depicted: transformation of light and shadow, perspective transformation and so on). In drawing practice, you might not pay too much consideration on all those techniques; even for the huge tree in Fig. 1, it would be enough by well-expressing the branches on different layers and the texture of the trunk. Nonetheless, it's still important to be aware of those key points when recognizing or observing plants.

Once again, art of drawing comes from the outside. A landscape architect may draw trees as daily work.

Indeed it's necessary to spend some time figuring out the skills when learning.

注释：

　　本节所提到的"特殊"植物，其实就是指：最能体现场所特征、最具场所精神的那一类典型植物。这对表现一地的场所意向、意境极为关键。于故宫、太庙、天坛，就是古柏，若于颐和园、北海、景山等皇家园林而言，则是白皮松、油松。其他诸如留园的大枫杨，狮子林的大银杏等，都堪称一种承载着场所记忆的特殊植物。

　　本组图选自为故宫博物院所作的文物保护研究文本。图中的古柏完全按原样复制，艺术性和表现效果则更多体现在对整体色调的控制以及光影映衬和枝干质感的细致刻画等方面。右图是习作则带有较多的自由发挥的成分。这类古柏在北京太庙及中山公园红墙外现存数量较多，可作参考。细审之下，不难发现这些道劲倔强的枝干下暗藏的龙气、灵气，恰如若蒲柏老先生所说的英国农场里的那种特有的"地精"、"神灵"之气。

The "feature" plants mentioned in this section, in fact, refers to the typical plants which can reflect the site's context and spirit. It is critical to describe the site impression and atmosphere. They are the ancient cypresses in the Forbidden City, Imperial Ancestral temple, the Temple of Heaven; they are the white bungeanas and pines in such royal gardens as Summer Palace, Beihai and Jingshan; they are the big maple trees in Lingering Garden; they are the ginkgoes in Lion Grove. All of them are special reminders and symbols of a specific site.

Legends of the group are selected from the version that is drawn for the cultural wealth preservation and research of the National Palace Museum. Cypress in the figure was copied exactly as the original edition. Aesthetics effect results from the general hue of the picture, the shadow-light effect and the detailed descriptions on the texture of trunks. Picture on the right side is a draft with a lot of freedom in the drawing. There are a large number of existing ancient cypresses living in the Imperial Ancestral Temple and outside the red wall of Zhongshan Park in Beijing to be observed as references. Through careful observation, the hidden "dragon spirit" and aura of this tree can be found in the energetical and sturbborn truncks, similar as the unique 'Goblin' and 'gods' hidden in an old British farm referred by Mr. Pope.

注释：
本组图均为新青年网上教学音频录制时现场所作范图，分别为了阐述植物表现的色彩、质感、层次等几个方面的问题。上图为两片典型的松林，以多层花树和阔叶树层层掩映、分出远近。前景松林只画树干，线条和用色都极简约，以纯黑作松枝、叶片，犹如给画面加上一镜框，使前后层次分明。由于是快速范图，有些细节，诸如地面色彩和远景的松树枝干没有作进一步调整，层次上稍有零乱，可以通过进一步刻画，使之完善。

These four figures are the demonstration drawings I made on site while e-teaching on the New Young net. They explain several issues in the expression of trees, such as colors, texture and layers. The above picture shows two typical pine forests covered by layers of flowers and broad-leaved trees to describe the depth. The front side pines are only drawn with trunks and simple clean lines and colors. The pure black twigs and leaves frame the whole picture. Because they are drawn quickly as draft, details like the groundcover color and distant pines have not been modified yet and the layers are slightly mixed. It can be refined in the next step.

注释：
右图分别从林中和林子外面描绘了树林的特征，林中的层次由色彩对比而来，形成渐渐深入的缘隧道的感觉。

The right two pictures depict the characteristics of the forest from the view inside and outside the woods respectively. Different layers in the forest were brought by the contrast in color to create the feeling of entering a tunnel gradually.

植物层次

树木的表达

枝干为主的常绿树表现，首先在于枝干穿插，其次是对大树树干的描绘。大体分两方面：

1. 枝干本身的机理，如松为片状，柏，尤其龙柏为丝状、线状，这是必定要首先表现的。就像用墨线表明的符号、标点一样，槐树多褶皱，松树多鳞片，古柏多游丝，这就是景观树木的符号，一画上去，特征就出来，四两拨千斤，非常适用。

2. 干上投影，同样是一个省时省力的细化方式，小枝和叶片在主干上的多种投影，能有效地突出树形、区分前后层次。一棵树的层次往往自由阴影衬出，同样一片树林的前后层次，也是以阴影衬托为主要方式。

大乔木的前端细节较多，尤其是一片林子的最前景树，细节往往多到杂乱的地步，所以要先稍作取舍、归拢。比如，本人采取的

Plane Plant

Expression of Trees

Note two aspects when expressing evergreens: the crossing of branches and the texture of big trunks. Pay attention to the following two points:

1. The texture of trunks, e.g. the lamellar pine trunks, and gossamer cypress (particular Sabina chinensis cv. Kaizuca) trunks. Just like the symbols or punctuations marked by ink lines, the texture should be expressed at once. The symbols of landscape plants, e.g. the wrinkles on locust trees, the lamellas on pines, and thin lines on antiquity cypresses, are highly representing the unique features of different plants. Grasping those features would make your drawing more expressive.

2. The projection on trunks. As another effective skill of expressing details, its essence is to notice the various projections of twigs and leaves on the trunk, which will effectively protrude the shape of trees and separate different layers. Layers of a single tree can be expressed by shadows; the same is true of the layers of a forest.

Since there are usually too many details on the front side of arbors, especially the very trees of a forest on the front view, whose rich details are even messy. Therefore, you'd better make some selection and assembling in advance. For example, my approach is to divide a broad-leaved tree into three parts: the crown, the trunk, and the projection between the crown and the trunk. Define these three parts at first, you'll find the rest details are no more

方法是将一棵阔叶大树分成三块：树冠、树干及冠干之间的大投影。第一时间首先确定这三部分，以后多种详细刻画都相当于在三部分之间游走。

树冠下的主干往往是色彩最沉深，分量质感最突出的地方。有时可以直接用墨线涂出来或连着树冠暗部做排线处理。大树后端往往涂浅色，对比减弱，以便给后一棵树留下表现空间。

若是赏枝干型的大树则需要对枝干、叶片分别细化，并主要用投影分出主次枝条及前后顺序。这种大树往往千枝万条，无论写生还是画景观树，都需要加以归拢、细化。一般重点刻画最大的或出枝最前端的一两条，后面的千万枝条大多可与背景一起做排线处理。

画植物的一种类似钉头状的排线，用于表现丛林杂树效果很好，本人曾在多种场合示范过，网上有专门为学生录制的此类视频，可加以参考。

than wandering among the three parts.

The part of major trunk under the shadow of tree crown is usually the deepest in color and the heaviest in tactile sense. Sometimes we can fill it directly by ink lines or approach the dark area which is adjacent to the tree crown by hatching. As to the back part of a tree, we should use light color to reduce the contrast and thus leave space for expressing the tree behind.

For trunks-ornamental arbors, you need to specify the details of branches and leaves respectively, and separate the major and subordinate branches and their orders by casting shadows. Since there are almost thousands of millions of branches on a tree, you'd better specify and assemble them consciously, regardless in sketch or in landscape drawing. Usually, we concentrate on portraying the biggest branch or the branches on the very front, and hatching the rest thousands of millions of branches together with background.

I used to demonstrate on many occasions a pin-head shaped hatching, which is very effective in expressing various random trees. If you interested, you may find and refer to such kind of videos for students on the internet.

■ 石头的画法　The Depiction of Rocks

注释：

本组图山石景观，均为快速表达的现场范图，突出快速达意。基本原则是：1.色突出对比，如左图山岩上部的浅黄和山间云雾的深蓝形成强烈对比；2.用单色（黑或深灰）调和各个突出的层次，画面中部的意象化的松树林几乎黑色，起到了这一作用。在较大尺度的山水表现中，这两种方法往往交替使用。

在这种尺度上的细致刻画主要是针对"山"的形态和机理，而对石头本身的质感，一般不作过细的处理，避免画面混乱。

These four pictures of rocks and hills are also quickly drawn as demonstration drafts. The basic rules of this kind of quick sketch are: 1. Strong contrast in color. For instance, the light yellow on the top of rock contrasts against the dark blue of the fog among mountains. 2. Use monochrome (black or dark gray) to reconcile all the conspicuous layers. In this picture, the simplified pine trees in the middle shot are painted near black, which exactly played that role. In a large-scale landscape drawing, these two techniques are usually used in turn.

The detailed description in this kind of scale is mainly resigned to the shape and texture of the mountain. The rock texture is often omitted to avoid massy of the picture.

注释：

上图为庭院石景的表达，重点刻画石头的质感和重量感。对于一些点景大石的细致刻画有些特殊技法，比如基层的线稿宜刚，稍作顿折；第一层色其实最好用彩铅排线作固有色，用深色的马克笔画阴面，同色马克笔统一画面，凡是底层彩铅较重处，马克笔无法渗入，则成亮色，没有彩铅或彩铅较薄处，则以马克笔显色，如此深浅交替两遍，石材的质、纹都可以表达得十分完美，加上对局部的苔藓、裂纹的刻画，表达自然会比较完善。

These two pictures are expressing the stones in garden. The texture and weight of stones are the main focus. There are some special techniques for drawing the big decorating stones in a site. For example, the hatching of rocks should show definition and hardness and some breaks in brushwork; the first layer is better to be in color pencils to paint the original hue of rocks, then use dark marker on shadow side; the same marker as before to unify the whole picture; marker color turns to be bright where the color pencils were used intensely and to be quite dark to show the color of marker where the color pencil hatching is not so dense. With these two layers of light and dark, both the texture and weight of the stones can be expressed perfectly. Adding some depictions on mosses and cracks, the picture will be more natural and vivid.

■ 水、天空的画法

■ The Depiction of Water and the Sky

无论是建筑图还是景观图中，水和天空的画法都宜简洁明快，用色彩以块面表达为主。高试点、大鸟瞰的天空宜画得较平，但用笔要灵活快速，多用线状笔触，一般三笔两笔就能成图，务必表现出天空的深远感和云气流动的意象。

天空用色完全视表达效果和环境感受而定，未必要一律用蓝色。事实上，黄色、橘色，甚至玫瑰色都可用以表达不同氛围的天空。需要注意的是，许多初学者在表达天空时常常画过头，虽然天空只是画面背景，但由于面积大，对画面整体性有一定影响，天空的表达应服从画面总体的色调。

水面的表达基本服从于周边环境需要，一般大鸟瞰的水面与天空保持一致，稍加岸线投影，点缀些许船只、帆点或喷泉即可。人视点的水面极复杂，近景的描绘以倒影为主，用笔宜活；中、远景则不要拘泥于亦步亦趋的倒影，可以适当虚化、留空，有时稍稍加入天空的色彩，感觉会更好。

The water and sky shall be expressed in simple bright colors by flakes in either architectural or landscape presentation. The perspective of sky in high position with a bird's-eye view shall be more flat by flexible lines. Basically, it can be completed in two or three quick strokes, but the sense of high and distant sky and the image of flowing air should be well expressed.

Decide the color of the sky according to the expression effect and environmental sense, unnecessarily monotone blue. As a matter of fact, we can use yellow, orange, and even rose to express the sky of different atmospheres. Though the sky is no more than the background, but its large scale on the picture will affect the whole presentation anyway. Since many beginners often over-render the sky, notice that the expression of sky should submit to the overall tone of the whole picture.

The expression of water also depends on the

surrounding environment. Generally speaking, the water would be in accordance with the sky in ordinary high position bird's-eye view; just adding shoreline projections and decorating with boats, sails, or fountains would be enough. However, the water surface in human's viewpoint is extremely complicated. The expression of water focuses on flexible portraying of reflections in nearby view. Unrestricted to the mirror reflection, the middle or distant view would be more expressive by proper blurring or leaving blanks, being more satisfying when sometimes the color of the sky added.

注释：
　　本组图天空局部均为快图常用方式，彩铅一抹而过，局部稍加点缀，条状用笔为主，点到即止。即使是大规模精细渲染图，天空的表达也不宜过分强化，宜干净利落，忌反复涂抹。

These three pictures of partial sky show the commonly used technique in quick sketch: slightly use of color pencil with little embellishment in details somewhere, strip brushwork, the less the better. Even in a large-scale meticulous rendering, the sky should not be emphasized too much. It is better to be neat and clean. Repeatedly drawing on the profiles is the last thing you should do.

荒野中的云（晴天）　Cloud in wilderness in a sunny day

荒野中的云（阴天 - 乱云飞渡）　Splashed cloud in wilderness in an overcast day

■ 水、溪流的画法
The Depiction of Water and Streams

Venis Water Lane From Grate Channel.

注释：
　　池塘宜静，水面宜平，用色平滑统一，稍加环境色即可。

　　溪流、叠水宜动，光影丰富而灵动，用笔也相应灵活。飞白光影一般可留空，在快速图中也可以先满涂水色，然后直接用涂改液点出溪流的飞白，这种方式与水彩画中用白粉点出光影的道理完全一样。

Pond should be quiet and the water should be flat. Color it smoothly with a uniform hue and only add a little color to the environment.

Streams and cascading should be dynamic. Light and shadow of them should be rich and vivid. Flexible brushwork is required accordingly. Highlight can be left as blank in the figure or can be quickly applied with full water color and then point out the flying white by direct use of correction fluid. This is exactly the same as white powder used in watercolor to point out the highlight.

Stone Brook
From Sargent Work

9. 石溪跌泉·叮咚山
（源流探险区）

■ 建筑及城市意象的画法

■ The Depiction of Architectures and Urban Imagery

这里的建筑是专指景观表现图中的建筑配景，与真正意义上独立的建筑画艺术不是一个层面的概念，后者有完善的思想、概念，往往表现出更严格的建筑空间诉求。本节所说的建筑及城市意象的概念相对较小，其核心词是配景。无论它以城市街区形式出现，还是作为大规模公园绿地的背景，它所追求的往往是城市与绿地相互渗透的概念。

建筑多以组群形式出现，前景有大片绿色或临水开阔空间，与正式意义上的建筑表现图相比，这种组群建筑表现一般比较简洁，省略了建筑图中必须的窗、幕墙等细节，只留下大的形体和色彩意象，虽然寥寥几笔，但用笔十分肯定，在反映城市意象的特大场景中，建筑立面上往往只保留两种基本元素和色块，一是明暗，一是投影，

"Architecture" in this book specifically means the architecture as entourage in landscape presentation. In fact, it differs from the independent art of architecture drawing which has its own complete system of thoughts and conception, and the architecture in landscape are usually more restrictive on space expression. In this section, architecture and urban image limits to the objective view with entourage as the key word. No matter it serves as the city blocks or the background of green space in large scale parks, it always seeks for the interpenetration between urban and green space.

Architecture often appears in building clusters with large green space or open waterfront space in front view. Compares with the architectural presentation in formal meaning, this kind of building clusters are often concisely expressed by few simple but definite strokes, only presenting the huge form and its color image while ignoring the necessary details such as windows and curtain wall. In extremely large scale scenery reflecting the urban imagery, there are only two elements left on the elevation: the chiaroscuro, and the cast shadow. Illustrators would often concern more about the cast shadow on the

注释：
左图节选自河北和山东两处城市设计项目的城市意象表达，共同特点是城市楼群与带状公园均面临大水面一字排开，视点较平，很容易画成平面展开的样式，而使空间大尺度的进深感无法体现出来。解决的方法是强化立面上的楼群投影，即在建筑立面上画出周边楼群的投影。这一步需要用笔快、准、狠。快，即用笔迅速干脆，稍缓则凝滞；准，即落影合乎逻辑，透视准确；狠，指敢用重色，准确的重色、宽笔横扫，往往可以一锤定音，但使用不当，也很难弥补修正。这方面对本人手眼配合及对工具性能的熟练程度有一定要求。

加强进深感的另一个方法是强化街道和行道树的透视缩进效果，街道形成的通廊可以很巧妙地引导视线，将城市一直引向灭点天际线上。强化方式是：1.由近至远打排线；2.楼宇落影由近至远加密。两种方法并用可以有效提示空间进深感。

These two pictures are selected from the city planning conception of Shandong and Hebei Province respectively. The common point in them is the city buildings and parks are lining up facing big water. The view points are rather lower so the perspective is easily put as a long scroll without screen depth. Solution for this problem is to stress the shadows of the surroundings projected on building facades. This step needs the brushwork quick, accurate and definite. Quick means the pen moves fast without hesitation to create a clean facade. Accurate means the shadows are drawn right logically. Definite means the dare to use dark and heavy color bravely. Big brush with dark color can usually determine the picture and is also hard to be modified when misused. This technique requires the good coordination between eyes and hands and the familiarity with painting tools. Another way to strengthen the sense of depth in screen is to enhance the perspective and indentation effect of the street and the trees along them. The liner space formed by streets has a good role in guiding the reader's attention to the distant skyline. To strengthen the perspective sense, first you can hatch from near to far away, and then put denser building shadows from near to far. The sense of depth can be effectively increased when both methods are used simultaneously.

而以投影更为重要。一般本人多关注建筑立面上的投影，尤其在近景处，这投影类色块会下手很重，笔触做得非常对实，目的也是为了突出建筑在特定的城市空间中的落位和层次。

其次要注意楼群之间色彩的协调渐变，如由近景的暖色到远景的冷灰，由底楼的深棕色到楼顶的亮黄色，此类单色渐变、多色渐变以及用马克笔和彩铅交替变换退晕等组合的方式非常多，可视图面复杂程度以及场景色彩要求选择使用。经验丰富的本人往往会根据实际情况选用一至两种自己最为熟悉的退晕手法，逐步形成自己独特的表达方式。

本人的习惯做法是：在以城市意象为主要表达目标的图纸上，

elevation of buildings, especially in nearby view, and often use steady brushwork to present the color lumps of projection for highlighting the location and layers of architecture in certain urban space.

Also pay attention to the coordination and gradient of colors among buildings, e.g. the change from warm color in nearby view to cold grey in distant view, and dark brown of ground floor to bright yellow of top floor. Decide the color changing mode, such as monotone colors change, multiply colors change, or grouped change by using markers and color pencils alternatively, according to the complexity of the picture and the colors of the scenery. A proficient illustrator would often focus on one or two most familiar color changing modes according to the practice, and gradually form his unique style by practicing the skills.

My usual practice is, to use multiply colors for combined gradient change and finally assemble all the colors by black or dark blue (hardly be recognized) shadow. In the urban background which mainly presents the green space, usually I choose

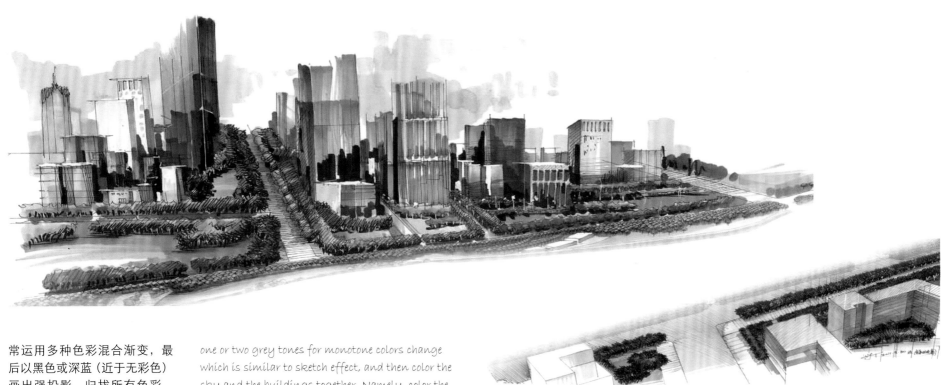

常运用多种色彩混合渐变，最后以黑色或深蓝（近于无彩色）画出强投影，归拢所有色彩。在表达绿地为主的城市背景中，一般仅仅选用一至二种灰调作近于素描效果的单色渐变，最后将天空和楼群一并着色，也就是把环境色作为建筑的统一意象色彩，起到背景作用即可。当然这只是一种个性化表达方式而已，读者完全不用拘泥和亦步亦趋地学习，事实上除了图例中展示的这些画面以外，本人平日所作范图选用的色彩和笔法是极为丰富的，限于篇幅，无论是文字和画面都不可能一一详加说明，读者在平日作图中可根据这些范图，多多加以练习，自然可以举一反三，形成自己的风格。

one or two grey tones for monotone colors change which is similar to sketch effect, and then color the sky and the buildings together. Namely, color the architecture with the same colors of the environment to create a uniform imagery, and thus contributing to the unification of the background. Since this is my personal way of expressing architecture, you don't have to restrict to it or follow it step by step. In fact, besides the illustrations in this book, I had demonstrated various ways of expression by various colors, due to limited space where I can't depict all in details. However, I wish you can draw inferences from those illustrations, and I believe you can form your own style by daily practice.

注释：

芝加哥意湖滨，选择了芝加哥最具代表性的区域——沿密歇根湖滨的格兰特公园及周边的高层建筑群作为城市意象的代表。右图为芝加哥白金汉大喷泉及其周边楼群组成的标志性城市意象，最大限度地利用光影塑造远景城市的变化和层次。

Chicago Lakeside. This picture chose the Grant Park along Lake Michigan and the surrounding skyscrapers as the most representative urban image of Chicago. The right picture shows the landmark city imagery formed by Chicago Buckingham Fountain and the high-rise buildings around it. The effect of light and shadow are maximized to express the distant view of urban scene.

注释：
　　哈尔滨城市意象：一组速写形成对一个城市意象的连贯表述。

　　这种做法类似连环画，用一组图反映一连串的城市空间意象。类似的尝试在建筑和城市规划领域应用较多，景观专业表现图中其实也可以采用这种连续拼贴的表达方式。老一辈建筑师彭一刚曾使用这种连续拼贴的方法表达中国古典园林的空间特点，取得了良好的效果。他的《中国古典园林分析》一书是这方面的杰作，全部使用单色线描、排线描绘，这种方式很值得学习和继承。

　　这类空间表达，本人尝试较晚，没有成熟作品可供参照，仅稍作提示。

City imagery of Harbin: a serious of sketches to document a coherent imagery of a city.

This express approach is similar to the comic that put the urban imagery as a serious of pictures. This approach can also be found in urban planning and architecture expression. And the collage of continuous images can also be used in landscape. Chinese senior architect Mr. Peng Yigang had employed this approach to express the space characteristics of Chinese classical gardens and achieved good results. The book "Analysis of Chinese Classical Gardens" by him is a masterpiece in this area. All the pictures are drawn in monochrome lines. That is quite worth studying and inheritance.

I have tried this expression approach not until recently. So there is no reference mature works but a little hint on it.

■ 配景 | 车、船小品
Ornamental Scene: Vehicles and Boats

注释：
　　车船一类的小品，在空间中尺度较小，往往是景观的点缀，但此类元素多居于前景，位置重要，若处理不当，非但不能为画面增色，反而会破坏图纸的艺术效果。
　　处理的基本原则是，用笔宜简省，色彩要鲜亮，下笔要肯定，细节要省略。由于是点景之笔，比例、透视任意一点错误都很容易被看到，所以对比例、透视要求要相当准确。一般可以整块地画，诸如把手、车牌、司机、舵手、小帆等细节，只要与特征表达无关，均予以省略。

Elements like cars and ships often play the role of decoration in the landscape in small scale pictures. But this type of elements usually appears at an important position in the foreground of a picture. So if they are not handled properly, they not only can not adorn the picture but to destroy the result of drawing.

Principles to handle them are using less lines, bright colors, unhesitating brushwork and omitting most of details. As the critical element in a picture, any mistakes, like perspective or scale, can be found easily. So it requires quite accuracy in proportion and perspective. Usually it can be drawn as a whole and the details like handles, license plates, drivers, helmsman, sail and etc. can be omitted, as long as they are irrelevant to the specific character to be expressed.

注释：
　　各种透视角度下的车辆。

　　在常规景观视图中，平行透视使用较多。在这种类似正视图的透视中，尤其应注意车身与两轮之间的相对位置。正透视轿车的前轮轮毂为一线，几乎看不出椭圆形，后轮轮胎宜放大，比实际尺度稍宽，则显出稳固。无论何种角度的车辆都应画出完善的投影、务必使车辆"落地"。

Vehicles from different perspective angles.

In the conventional landscape plan, lower-right parallel perspective is often used. In such a perspective from the front, extra attention should be paid to the car body and the relative position between the two wheels. Viewed from the frontside, the car's front wheels are as a line. There is almost no oval showing. The rear tires should be enlarged slightly wider than the actual scale to show solidity. And no matter what angle is taken, the vehicle should project logical shadows to make sure the "landing".

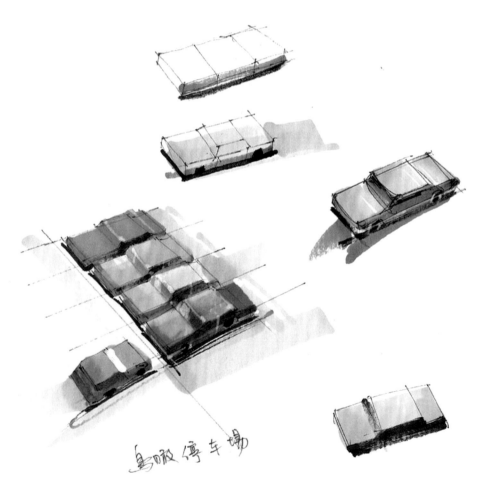

注释：

正侧面车辆，注意弧形门线，底线和前后保险杠位置，车身着色时，应将其想象成筒子型，车门弧形线正好是筒形的结构线，通常画法是用细涂改液在车身上一笔勾出。基本形是先缓缓地外扩至底盘处急收，然后稍作停顿，形成门线下的一个亮点。对于正侧面车辆而言，这根白线的准确度，关系到整个车的透视，平时宜多加练习。

车身着色用笔，本人称之为"刷"，一般正面一笔，侧面一笔，底线连前后杠为第三笔。一笔比一笔深，三笔成形，然后作车窗及高光、投影等细节，务求笔法干净利落。

Viewed from the front or side sides, the curved door lines and the bottom line, the position of front and rear bumpers should be paid attention. When you color the body of a car, treat it like a tube. Then the curved door line is the structure line of a tube. The normal method is drawing it at one stroke with the correction fluid. The basic shape is to extend slowly until the sharp stop at bottom, then to stop there a while to create a highlight spot at the bottom of the line. For a car from this view, the accuracy of this line matters dramatically. It is advised to practice more.

The brushwork when you color the body of a car, is called "brush" by me. Generally, there is one stroke for the front side, one for the side face and the last for the bottom line together with front and rear bumpers. They are thicker and deeper one by one to form the shape. Details like windows, highlights and shadows are then to be finished by clean and neat brushwork.

■ 人物的画法

■ The Depiction of Human Figures

人物是景观图中的点睛之笔，对活跃气氛、标示空间尺度和空间功能有很大作用。在常规的大场景鸟瞰图中，人物多用简笔、单色，落影即可，无需多加修饰，但在透视图中，近景人物需稍作刻画，基本原则还是用笔简省，但要突出人物的动态。这种写意人物大多三两笔即成形，动态变化强调适度，合理。

表现动作较大的人物应注意：

1. 无论何种肢体动作，重心均落于两脚之间或承重脚上。

2. 人物的肩线、乳线和胯骨线三线永远处于相逆的方向，否则人物必然倾倒。

As an element that brings a vivid presentation to life, people play an important role of crowning touch in conveying an active atmosphere and indicating the scale and features of space. In ordinary large scale aerial view, people are usually expressed by simple touch or shadow in single color; too much embellishment is unnecessary, as long as their image was displayed on the picture. However, in nearby perspective view, you'd better slightly portray the people. The basic principle is still to be simple and concise, while highlighting the motion of people. Often in two or three strokes can the illustrative people be drawn, while proper and rational motion changes are emphasized.

Note two points when express people making dramatic action:

1. Whatever gestures they made, setting their center of gravity between their two legs or on the bearing foot.

2. The three lines of people, namely, the shoulder line, the breast line and the hipbone line, should always be on the opposite direction, or the people would fall down.

注释：

　　这种貌似复杂的场景，其实均为单线平涂，速度极快。油性马克笔在小面积作图上的渗透、叠加方面的优势展现无疑。不用一笔水彩，足以体现水彩的韵味，读者有兴趣可稍作尝试，必定会有所收获。

　　特色场景，特殊动作下的人物，注意视点的变化。

This seemingly complex scene, in fact, is flatly coated by monochrome in a very short time. When painting on a small screen, the benefit of Oil-based marker is significant with the penetration and overlapping effects. It can produce the effect of water color without using any a little water color. Try it if you are interested and you will not be disappointed.

These are people features with special poses in a special scene. Pay attention to the change in view points.

第二部分

CHAPTER TWO

过程图解析

■ 过程平面图

关于一幅效果图的创作过程或作图程序，在这里提几点个人意见。

首先，传统的建筑渲染图的创作过程可能已经不适用于今天以快速表达为主的表现图作业。那种先敷底色，然后逐层叠加，逐步完善的作图方式，在 20 世纪八九十年代的环艺表达中曾普遍使用，对应的作图工具是水彩、水粉和专业的覆膜喷绘。本人曾有十年左右的时间，大量采用喷笔加马克笔的混合画法，创作过数以千计的环艺图纸。但进入以快速表达为主的时代，本人认为这种创作方式和作图步骤也不应该一成不变。

事实上，近几年来本人所作大量渲染图均是在数小时以内完成的快速作品，传统技法上的先浅后深，先大面后细节等程序几乎没有得到过应用，换句话讲，本人的快图作品几乎都是一遍上色完成，很少出现多层叠加覆盖的现象。程序大体如下：

Analysis of Process Drawing

■ Process Plan

I don't have too much to comment but I wish to give some personal advice on the creation or drawing process of a presentation.

First, the creating process of traditional architectural rendering may not catch up the steps of current presentation which is dominated by quick expression. The drawing process of spreading background color initially, then overlying layer by layer and finally gradually completing the picture, was once widely adopted in environmental art presentation in the 1980s to 1990s, with the counterpart drawing tools are watercolor, gouache, and professional coated inkjet. I used to create thousands of environmental art drawings by using airbrush and markers jointly for 10 years in the past. However, in the age of quick expression taking the lead, I think this way of creation would not and could not be unchanging evermore.

As a matter of fact, lots of my renderings in the

past few years are finished within very short period. Those traditional skills on process, such as light colors prior to dark colors and general scene prior to the details, are hardly applied. In other words, there's seldom multilayer overlying and covering in my works; my quick presentations are almost completed by coloring once. Following are my steps:

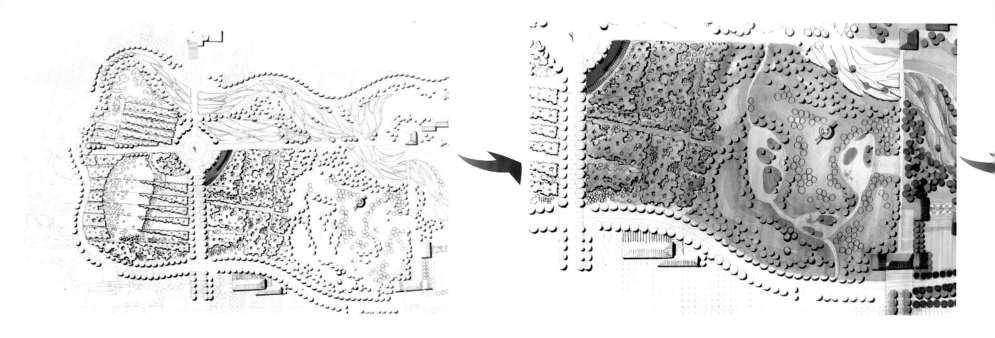
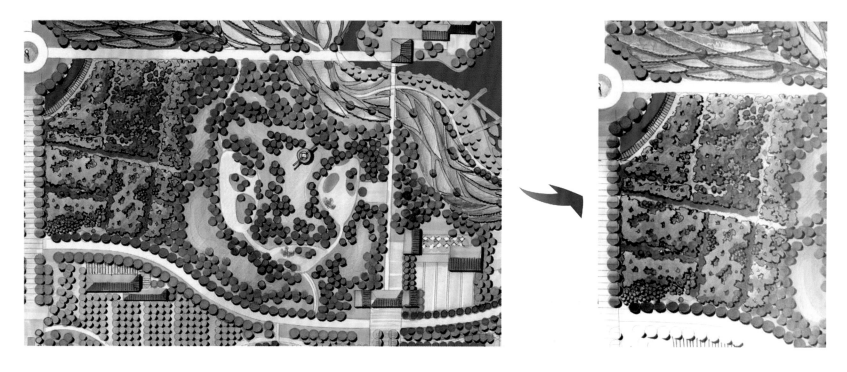

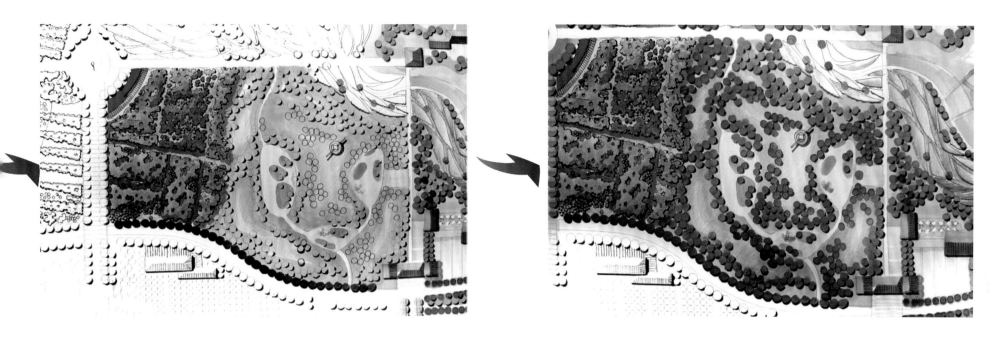

1. 完善墨线图。通常会在 Sketch Up 线图基础上加绘很多近景细节，包括细化近景的树木，加画大量排线，分出画面前后进深层次。这一步相当重要，后期图纸上许多细节刻画都要以这种大面积排线所设定的层次进行，这一步底色的铺色水平与本人的艺术素养和空间形象表达能力有很大关系，几乎没有什么智能软件可以替你完成，需要勤加练习。

2. 按一定顺序（从左到右，从上到下，或者彻底反过来）上色，大部分画面上色均要求一遍完成，马克笔用笔特别强调一步到位，繁复叠加的画面既容易脏又易凌乱琐碎。本人的方式大体上是先刷后点，以刷为主，明暗退晕大多是由落笔的轻重和笔在纸上停留的时间决定，看似一笔扫过，但轻重缓急都在其中，这是画草图过程中很享受的一个步骤。极少数重点部位会先铺彩铅底色，借助彩铅底图的低渗透性，实现较大面积的平缓的退晕过渡，对于创作大面积的草坡、广场特别适用。

1.Complete the ink line drawing. I usually add more details in nearby view on the sketch drawn with Sketch Up, such as refining trees in nearby view, adding plenty hatching and separating the layers. This is very important since many detail depicts on the following drafts are added according to the layers set by this large area hatching. The proficiency of spreading the background color depends on the artistic quality and space expressive ability of the illustrator, hardly any intelligent software could help you except your hardworking.

2.Color in certain order. (It doesn't matter if you color from left to right or from up to down, or reversed.) Since too much overlying and covering would easily make your drawing looks dirty and messy, most of the coloring should be finished once, especially when you use markers. Generally speaking, I approach mainly by painting, and then stippling. The color changes from brightness to darkness are mostly decided by your strength on the touch and the timing of your pen point on the paper. Controlling all the rhythm by a single touch, what an enjoyable moment in sketching! Seldom important parts are required to spread the background by color pencils at priority, for the low permeability of color pencil effect are perfect for large area grassland or plaza to achieve a smooth color change.

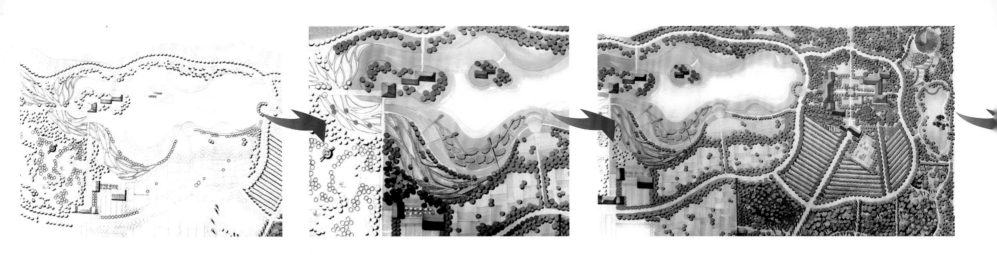
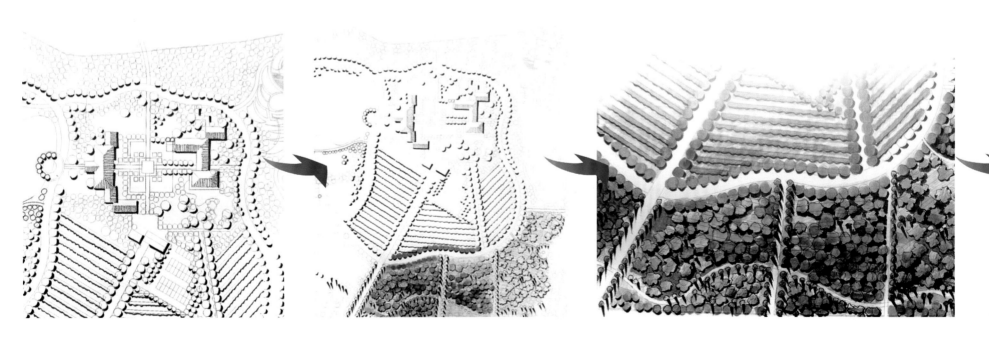

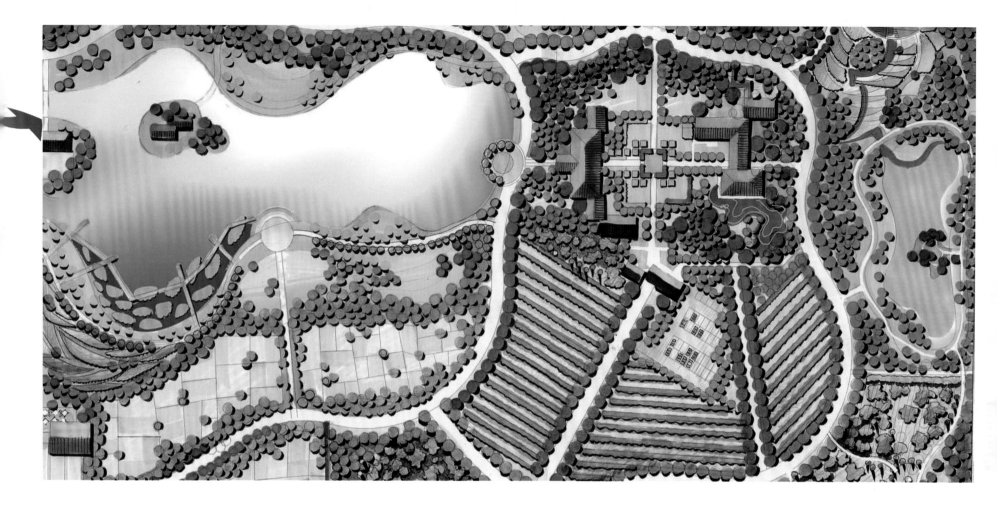

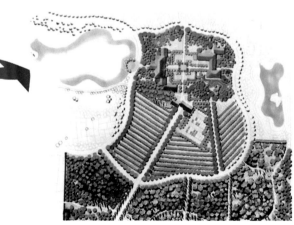

3. 对于大尺度场景而言，基本铺色完成后，会使用墨线排线将画面稍作归拢，这是统一的一步；然后用涂改液点出细节元素，这是对比的一步；最后用一支粗墨线笔，将画面中所有的投影一次完善，并在重点部分用墨线排线再次整理一遍，即可完成。

以上三步在许多局部透视图中可能会进一步简化。

3. For large scale scenery, when the general ground coloring has been completed, hatching in ink lines to assemble the picture (the unifying step), then stippling the details by correction fluid (the contrasting step), and finally use a thick line pen to complete all the projection in the scene at once, and hatching again in ink lines to modify the important parts.

In many part perspective views, these three steps can be further simplified.

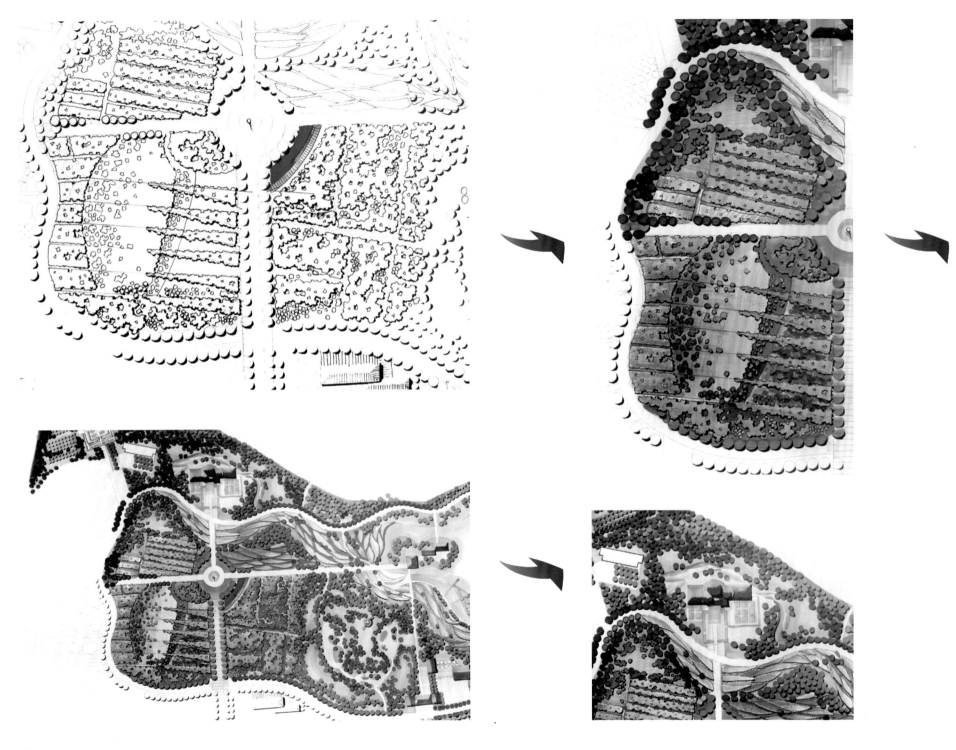

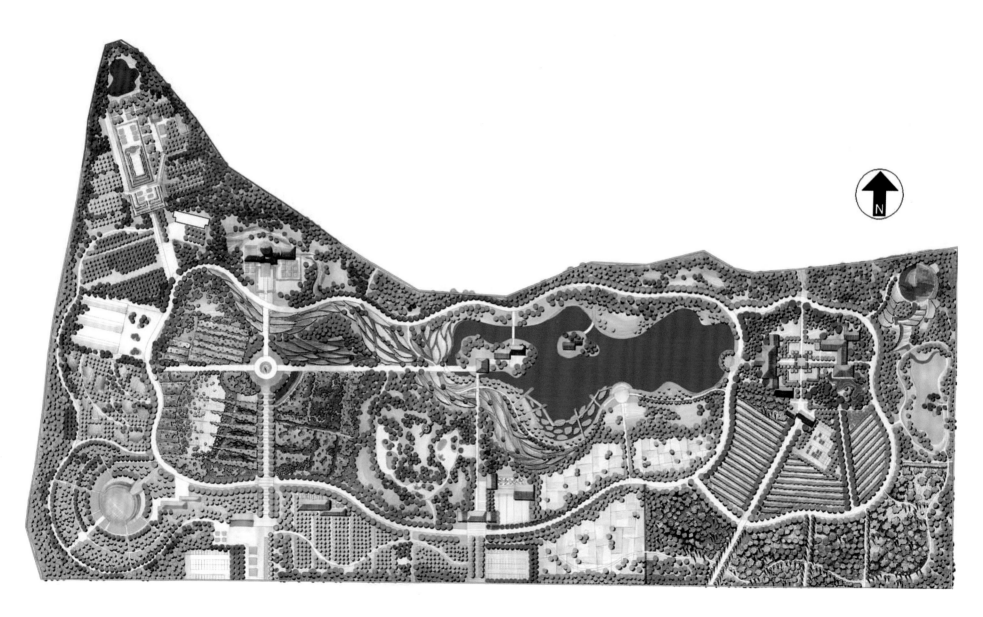

■ 过程鸟瞰图的画法   The Drawing of Aerial View

注释：
在 Sketch Up 线图基础上加画排线，完善植物布局和开放空间形式，加画码头、帆船等配景。

用彩铅和马克笔混用，画出渐变的底色。

点出片林、小路、建筑和帆船，用蓝色彩铅在湖面上作排线，渲染出前后层次。

Hatch on the Sketch Up draft to finish planting arrangement and the form of open space. Add accessories like dock and sailing boat.

Paint the background as gradient by the mixed use of color pencil and marker.

Draw out the woodlots, paths, building and the sailing boat. Paint the lake by blue color pencil to create the sense of front and back layers.

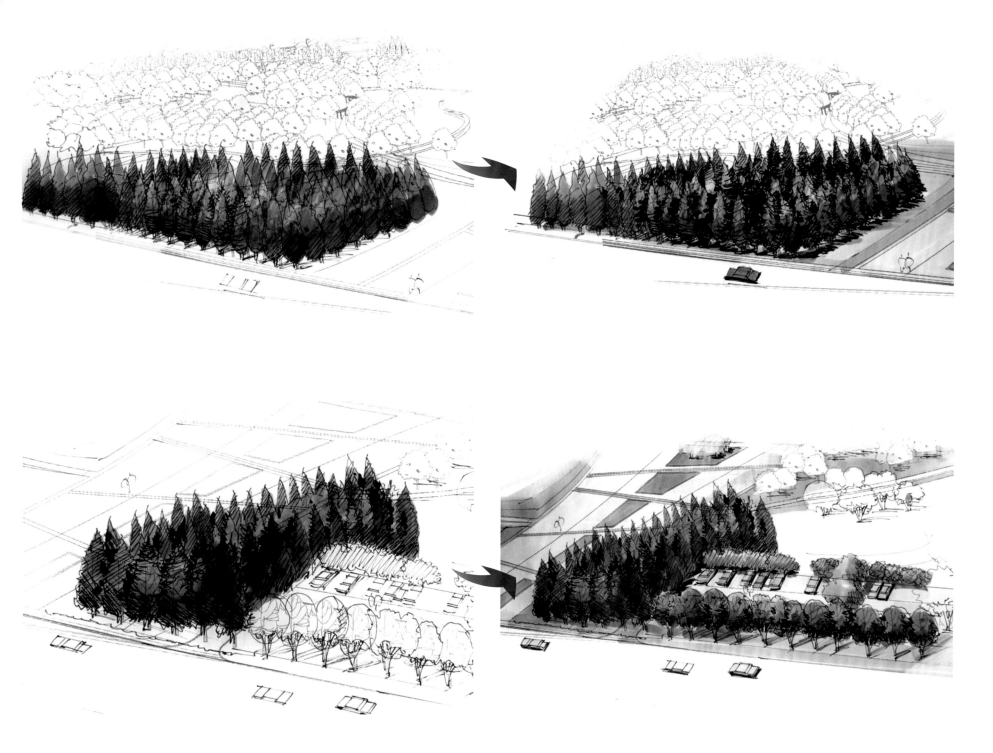

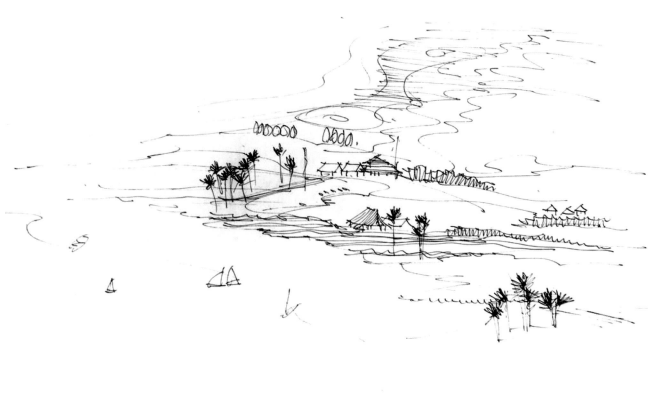

注释：
补齐线图中的注意事项：

特色植物（棕榈等）简易表达；
排线与投影；
月形水湾及沙丘内侧防护林的强调等。
元素均需在这一步落图，从这个意义上讲，补齐线图的过程也是设计方案细化和再次推敲的过程。

Notice for the step of improving line draft:

Express the feature plants (palm, etc.) simply;
Hatching and cast shadows;
Stress the moon-shape bay and the protection forest on the dune.
All the elements should be cleared in this step. In this sense, improving the line draft is a design refinement process.

注释：
　　下图是本人上文所述"一遍上色"的典型。整张图从线稿至上色完成用时不超过两小时。这类草图着色的目的主要不是为渲染环境，而是进一步说明场地的特征。一些重要元素，如沙丘的影子、外海与潟湖、防风林带、内湾的月丘等，通过着色都能更清晰地显现出来。这对理解方案及团队交流所起的作用往往是不可估量的，而且这种简易渲染省时省力，其效率比同尺度电脑渲染图快上百倍。全部着色时间约半小时，基本笔法是将所需的层次依次"刷"出来。

This is a demonstration of "coloring all in a breath" mentioned above. All the procedures from draft to coloring took less than two hours. The color used in this kind of draft is not for environment atmosphere but to illustrate the features of the site further. Some essential elements can be more significant and visible by colors, like the shadow of sand hill, seashore and the lagoon, protection forest belt and the Moon Hill within the bay. This will help to the understanding of design and the communication about it. It would not take long and much effort to finish this kind of rendering. So it is much more effective than a computer render. All the coloring work took about half an hour and the basic technique is to "brush" out what you want layer by layer.

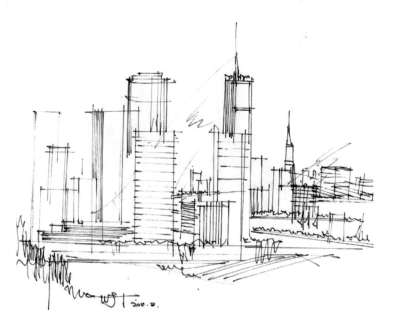 

■ 过程建筑的画法

■ The Drawing of Architectures

  从广义的建筑学角度上讲，景观、城市规划和建筑是一体的，当前中国，大尺度的城市设计项目往往被划分为单纯的规划、建筑、景观园林三部分工作。在景观规划中，加入对城市空间的考量和部分城市风貌控制在许多景观团队中已有不错的实践，就表现方式而言，本人倾向于把这类混合风貌的表现图分为以城市为主和以景观为主两类。

  两类图纸在布局、落影、用色等技法上有很大差别，本书着重介绍前一类城市风貌为主的表现图技法。建筑类城市意象图最直接的要求是直观清晰，所有技法围绕这一目的进行。楼间层次和相互间的落影至关重要。为此，需有意降低建筑固有色的影响。正如本节的两个案例，第一例为潍坊中心区高层楼群的意象表达，重在渲染气氛，以蓝色和金黄两色形成中央区最突出的对比，其余大面积深色阴影，用以调和整体色调。画面视觉冲击力很大，却能章法不乱，鲜而不俗。第二例为纽约天际线复原，基本维持原建筑群的组群形式和基本色调，只加归拢，并不人为强化。力图准确反映出世贸大楼、

Architecture, in a broad sense, includes landscape, urban design and architecture. Because of the historic separation of administrative responsibility in China, large scale urban design projects are often simply divided into three parts: planning, architecture and landscape architecture. However, I believe the integrated design will be the future. By adding urban space consideration and certain urban landscape control in landscape planning, good effect has been made in many landscape practices. I'd like to separate this complex landscape presentation into the kind that focuses on urban and the kind concentrates on landscape in ways of expression.

These two kinds of drawing are greatly distinct in the skills of layout, projection, coloration, etc. This book concentrates on introducing the skills of urban landscape presentation. Since the most direct require for this architectural urban imagery is to be straightforward and clear, all the skills are serving this purpose. The cast shadow between buildings as well as the overlapping shadow is so crucial that we should prevent its affect by the color of the architecture. Take the two cases in this section for example. One is the urban imagery presentation of high-rise buildings in CBD of Weifang city, which focuses on rendering the atmosphere, highlighting the CBD by the contrast of blue and gold while adjusting the overall tone by large dark shadow on the rest parts. This picture gives a great visual impact, unusually composed but not ridiculous, gaily-colored but not vulgar. Another example

注释：
1. 简易线图只画出对天际线最具影响力的建筑群及城市前景，后部大量楼群是在后期"点"出来的。
2. 第一遍色实际上是落影、分层，这一步最关键，几乎是"一锤定音"的步骤。影子永远是建筑物"落地"的重要步骤，阳光自城市上方射入，上浅下深的色彩布局自始至终都要保持好。
3. 加画中间层，即固有色层。前景为亮黄色，后景用灰蓝色，在色彩深浅、饱和度两方面都对图像进行了主观的人工处理。
4. 加画天空和近景的水面和船只，完善场景的层次，点出远景的建筑群。最后的处理过程宁简勿繁，点到为止，忌处处闪光、杂乱无章。

1. Simple line draft is only with the profile of the urban outlook and the most influential building groups to the skyline. Other buildings are "pointed out" in quantities in later procedure.
2. The first and critical color layer of paint is to distinguish layers of the screen by casting shadows. This step is so important that can almost determine the whole picture. Shadows are essential for the building to 'land' on the ground. Sunshine comes from the sky into the city. The gradient effect from light to dark is important and should be paid special attention to in the whole process.
3. Add the medium layer of color which is the inherent color of items. The foreground color is light yellow and background color is gray blue. Adjust the picture subjectively in the brightness and saturation of colors.
4. Add accessory elements like sky, nearby view water and ships. Improve the sense of layers in the picture. Point out distant buildings. Less is more in this procedure. You should avoid overdoing to make the picture flashing everywhere and being messy.

帝国大厦、自由女神像及巴特利公园沿线摩天楼群在纽约天际线上的真实尺度和基本光影效果。人为处理的成分较第一张明显减少。

在实际的城市设计项目中，对于现状街区和改造街区的表达原则大体应按照上述两例的不同方式进行。

is the restored skyline of New York. Generally maintaining the group form and basic tone of the former building clusters, slightly assembling but not deliberately intensifying, this picture aims to reflect the realistic scale and basic light and shadow effect of World Trade Center, Empire State Building, Statue of Liberty, and the skyscrapers along the Battery Park. Obviously, there are less deliberate touches than in the former case.
Generally speaking, you can refer to the two cases for the principle of expressing present blocks and reformed blocks in realistic urban design projects.

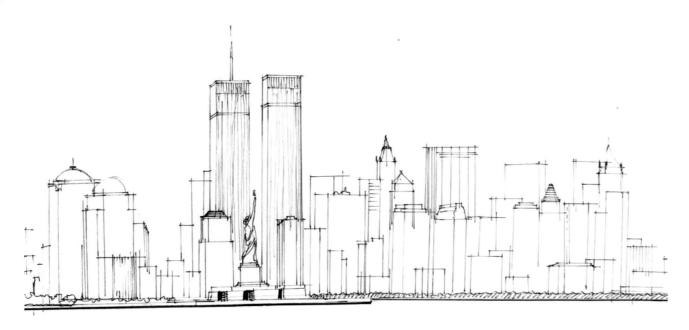

注释：
　　1. 落影、分层，画出作为背景的建筑群。
　　2. 基本色调宜一次到位，整个中部的楼群其实是一笔刷出来的，更关注整个楼群从上到下的光线色彩变化，各个建筑之间的微差变化是通过白色涂改液逐个调整完善的。
　　3. 加画天空、水面和前景的林带，完成画面。

1. Project shadows and put color in layers. Draw the background building groups.
2. Inherent colors should be finished at one time. All the buildings in middle layer were actually finished at only one stroke. Pay more attention to the gradient effect from top to bottom of buildings caused by light. The differences between each individual building were refined by white correction fluid later.
3. Add sky, water and the frontside forest to finish the drawing.

第三部分

CHAPTER THREE

项目说明

Project Description

本书的基本特征是"以例说法",将技法的解说融合于实际项目设计中。本章选取的设计大体是本人在以往三年中所经历,并作为主创设计师,针对项目的特征和个人对场地的理解所做的部分设计草图和表现成图。这种情况下做设计,有点类似于命题作文,和在无限制条件下,纯粹的"海阔天空"的艺术家创作是有所区别的,后者虽然创作的自由度较大,表现风格可以更为灵活,但是对于场地的气候、水土和落地区位等功能性方面的考虑明显不足,因而设计的适宜性也不会很出色,这是当前本人国景观设计的发展状况不能允许的。但这也是现时条件下,我们的院校教育中常用的"模拟性"、"研究性"作业容易犯下的毛病。所以,对于本章所列的案例,读者不如将其看成一种命题作业和限定性研究创作的产物。在确认多种限制的条件下,通过对多种信息的筛选、分析、循序渐进地制定针对每个项目的独特设计及适宜的表达方式。

第二个基本特征是限时作业。本书所选全部案例及作品都可以视为快速表现图,绝大部分图纸在 5 小时以内完成。通常笔者在方案确定以后,会在一天之内尽可能多地做出不同角度或不同功能区的意向图纸。事实表明,在完成平面向三维过渡的透视草图勾画过程中,会出现许多平面构思中未曾想过的问题,方案因此得到不小的改进。所以有时笔者戏称,笔者的许多方案是"画"出来的,草图在方案酝酿和具体化的过程中往往会起到很大的作用。但这一过

Taking illustration as its basic feature, this book interprets skills in realistic project designs. This chapter contains my selected designs in the past three years and part of design sketches and final presentations which were drawn according to the project features and my personal understanding of the site as chief designer. This conditional designing is more or less similar to proposition thesis, different from the artwork of free and unconstrained style which created in limitless environment. Though it may enjoy freer creativity and more flexible style in expression, for the obvious lacking of consideration on the features of the site such as climate, geography, and location, it wouldn't be remarkable in terms of adaptability of the design, which is almost unbearable in the development of landscape design in China. Unfortunately, our "simulative" and "researching" homework in present academic education often result in this kind of works. So I'd rather you take the cases in this section as a topic work or conditioned research. Confirm the restrictions so you can gradually develop the unique design and proper way of expression of each project by screening and analyzing.

The second point is limited time. All the works selected in this book can be taken as quick presentations, for most of which were finished within 5 hours. Usually, once the scheme was decided, I'll draw as many perspective views in different angles and of different function areas as possible in a day. A truth is that, when drawing 3D perspective sketches on the basis of plan, many problems that had never been considered in the plan would emerge and improve the scheme not a little for its sake. Sometimes I make fun that many of my schemes are just "drawn out", for the sketches usually play a big role in preparing and specifying the scheme. However, this process emphasizes the speed instead of the form. Regardless sketching purely by lines or rendering by single markers, drafting half or

程必须强调快速，不宜过分讲求形式。可以单纯线勾；可以单纯马克笔涂抹；可以只画半边、一角；也可以在多层草图纸上各画一层，相互替换比较。总之以自己最熟悉的方式不停地画。在这一过程中，你的思维一直会被牵着高速运转。所以，这些作品几乎全部用马克笔和彩铅混合表达，目的是体现快速、切题、表意的基本要求，特点是精准、简洁、巧妙、意到为止，绝不拖泥带水。为了实现快速达意的要求，有必要放弃一些对于画面效果"完善"方面的追求，尽可能将元素精简，越少越好。元素越少，画面识别性越强，可读性也就越强，这也是手绘强于照片，强于电脑的最大优势。手绘图融入了设计师个人的感悟、取舍、概括，呈现的是简化后，更明确、更醒目的整体，因而也常常意味深长，能够实现一种意象上更为真实的环境表达。手眼配合、心手合一的状态也是这么练就的。

a corner of the paper, or comparing and replacing different layers on multiply sketch papers, do the drawing in your most familiar way. In this process, leading by your hands, your mind is running at high speed. So you'll find these works are almost all presented by markers and color pencils jointly to answer the basic requirement of being fast, pertinent and expressive, and to embody the precise, simple, delicate and sophisticated but never sloppy features. To respond to the fast and expressive demand, we have to give up pursuing the "completed" effect; eliminate the elements as less as possible, for the less the elements, the more recognizable and readable for the picture. It's also the most important advantage of hand drawings comparing to photos or computer drawings: containing the personal understanding, selection and generalization of the designer while presenting a more accurate and more eye-catching entirety, the hand drawing often becomes meaningful, and further achieves a more realistic environment presentation in terms of imagery. Only though this process can we practice the state of hand-eye cooperation and mind-hand synchronization.

■ 河北清河县水系规划

Planning of the Qinghe River in Qinghe County, Hebei Province

市民中央公园鸟瞰　　Bird's-eye view to the public Central Park

注释：

河北省清河县清水河规划一草、二草平面图。

设计目标：优化滨水区环境，增强城市中心区活力；丰富滨水空间形式，聚拢人气；塑造规划新城区和原有建成区的两种独具特色的城市界面。

围绕设计目标，两轮设计均在工作草图阶段设计了多种城市活动开展的场所，以及大流量城市活动进入滨水区的多种路径选择：设计了完善的滨水游步道，以及与城市路网相连贯的快速路系统，由堤顶路、上层驳岸和滨水步道三层道路系统加以串联，形成综合的休闲网络，使大量人群能方便到达水滨，又可以快速疏散进入城市各大功能空间。在绿廊沿线设置多个商业和市民服务中心及体育公园等功能空间。在大型公共中心街区将滨水绿地直接引入城市，使之成为中心区重要的集散场所和市民中央公园。

The first and second draft of Qingshui River Planning in Qingshui County, Hebei Province.

Design goals: To optimize the waterfront environment, enhance vitality of urban areas; rich the forms of waterfront; attract more activities to this site; shape the different urban interface between newly-built district and existing area.

To achieve the design goals, two rounds of design modification focused on the versatile sites for public activities and the multi paths into this waterfront area. Design refines the walkway trail along waterfront and the high speed road system connected to urban road. A comprehensive network of leisure consists of the dike road, upper revetment and waterfront walk trail connected one by one. This network enables the crowd easily access to the waterfront and evacuate to other urban facilities quickly. Several commercial and service centers and functional facilities like sports parks are set up along the Green Corridor. And the water system is introduced into the urban area at a large public central block and turns the block into the Central Park, an important distribution center and public space of the city.

本人习惯的方式是在一张纸上多层涂抹，先墨线笔，后毛笔、马克笔，这一步加入的元素，想法最多，最后若错线太多，也不舍弃，改用白色涂改液修正，用粗墨线覆盖。事实上，本书中许多看似规整、漂亮的平面图和鸟瞰图，都曾经过多次涂抹、覆盖，这些不大清洁规整的图纸，往往是最好的设计催化剂，并且记录着设计过程中的点点滴滴。因此笔者认为，设计过程和表现过程有时是合二为一的，单单托着脑袋冥思苦想是出不来设计的，愿读者多多动笔，边画边想，这是真正的捷径。

I'm used to over-painting a paper, first by ink line pen, then by brushes and markers. In this process I add the most elements and ideas; even this may result in too many errors, I couldn't give up the drawing but modify it with white correcting fluid and then cover it with thick ink lines. In fact, many plans and aerial views which look clean and even beautiful in this book had been modified and covered for several times. Those unclean drafts are usually the best catalyst in design and the record of every drop in the process. So I think sometimes the designing process is combined with the expressing process; simply racking one's brain would end nowhere. I wish you to move your hands, to think while drawing: that's the real shortcut to successful drawing.

注释：
利用场地原有的河湾环境加以扩建，形成商业街区前端的开阔水面，成为鸟瞰新城的最优越的观赏点和城市风貌展示区。前景处放大的滨水硬质广场，为市民提供了多种滨水活动和交流的场所。

This project was developed from the enlargement of original waterfront. An open waterfront space is formed to be the end of a commercial street, which becomes the best site-seeing point to view the new town and exhibition space of the city. The enlarged hard-covered plaza on waterfront provides the public with the space for a variety of activities and communication.

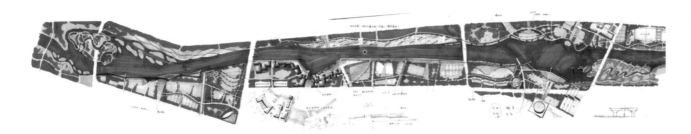

河北清河县水系规划平面图　　Planning of the Qinghe River in Qinghe County, Hebei Province

从商业街区鸟瞰的新城　　Aerial view to the new town form commercial district

唐山水处理厂景观设计
Landscape Design for the Water Treatment Plant in Tangshan

唐山水处理厂景观设计　　Landscape Design for the Water Treatment Plant in Tangshan

唐山水处理厂景观规划总平面图 Landscape design for the water treatment plant in Tangshan

镇中心博物馆人视点效果 Perspective of the town central museum from persons' view

黑龙江五大连池安置中心镇城镇中心博物馆广场

Central Museum Plaza in the New Settling Town of Wudalianchi, Heilongjiang Province

注释：
　　选取了场地一侧大面积的树阵作为表现主体，突出蔽护—安居这一概念，大面积树阵形成绿色的天棚，护佑着中心安置镇的市民，并将此作为日常休闲娱乐、城市交流的"舞台"和"后庭院"。

　A large tree matrix on the side of the site was chosen as the main depicting character. The conception of "shelter-settle down" was emphasized. The large matrix of trees formed a huge green providing shelter for the inhabitant in the center town and also the stage and backyard for entertainment and communication.

注释：
利用竖向高差，形成丰富的路径变化，景观也随着路径改变，步移景异。

Changes in altitude are utilized to create versatile paths. Landscape is also changed along with the changing path.

多层游步道系统效果图 Effect of multi-layer tour trail system

注释：
采用了半地下覆土建筑，有机形态设计（设计由清华大学建筑学院景观系完成，笔者只做其中表现图部分）。用笔较松，地面用笔模仿火山熔岩的形态，形成较丰富的意境，此为阶段性草图，在植物设计和空间围合等方面尚不完善。

A semi-underground building was designed with an organic shape (designed by the Department of Landscape Architecture, Tsinghua University and the author was in charge of the expression part). The brushwork was quite relaxed. Hatching on the groundcover imitated lava surface to depict a rich imagery. It is just a stage draft and still needs modifications on planting design and space enclosure.

镇中心博物馆广场鸟瞰　Aerial view to the museum in the town center

包头幸福广场
Happiness Plaza in Baotou

注释：
此广场为包头中心大绿廊上一个稍稍放大的节点，场地四面临城市机动车干道，南北跨多个街区，东西向需以步行道加以贯通。

设计要求：
1. 营造一个开阔敞亮的市民空间；
2. 完善场地的植物层；
3. 塑造具有地方文化特点的地标；
4. 南北两端需与原有城市绿廊有机结合，逐步融入城市绿色廊道之中。

This plaza is a slightly amplified landscape nod on the Central Green Corridor of Baotou. The site is surrounded by urban roads and covers several blocks from north to south. Walk path should be added through east-west latitude. Design requirements: first, create an open public space; second, improve the vegetation on the site; third, form the local landmark with special cultural feature; forth, link with the existing green corridor with the city at the south and north ending and it should be gradually integrated into the corridor.

标志区入口透视 Perspective from the entrance of symbolic zone

注释：
从景观大道上看入口标志，具有耐盐碱特性的东北黑松塑造了通往南区的视觉通廊。

View entrance sign from the scenic Avenue. Northeast black pine which is featured with salt-tolerance creates the visual gateway to southern zone.

标志区入口透视 Perspective from the entrance of symbolic zone

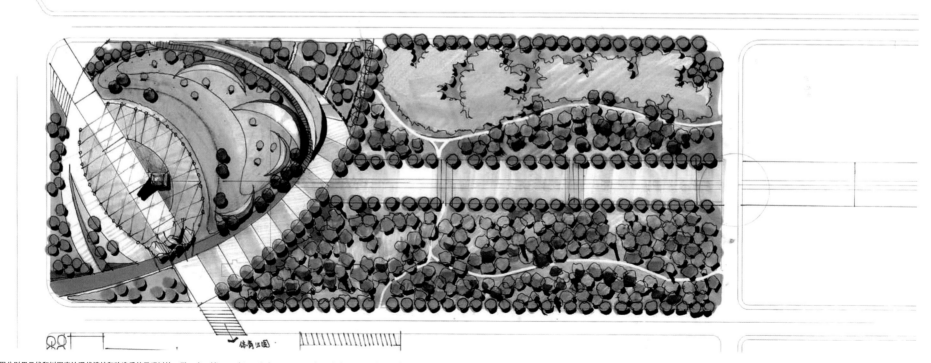

平面图分别用云线和树圈表达现状植被和改造后的景观树林。 The cloud line and tree circles represent the existing vegetation and recreated landscape forest respectively.

曹妃甸科技公园设计
Planning of Caofeidian Technology Park

注释：
　　鸟瞰图，用时1小时，A4大小，典型工作草图，设想了将原方案的大草坪改成观赏水池后的景观意向，及公园与新区楼层之间的呼应关系。技法基本平涂，能达意即可，点到为止，树群轮廓用涂改液画出。

This aerial view finished in an hour on an A4 size paper is a typical working draft. It depicts the landscape idea of replacing the lawn in original plan into an ornamental pool. The relationship between the park and buildings in the new district is also illustrated by it. Only basic hatching technique was used and it is quite enough to describe the idea. The profile of tree group was drawn with correction fluid.

从迁曹铁路方向看新区入口的"市民平台"　View from Qiancao Railway to the Public Platform at the entrance of the new district.

注释：
　　科技广场的方案鸟瞰之二，突出了方案专门设置的樱花大道纪念景观，以及公共草坪与城市道路（人行道）之间的关系。画面几乎不做任何修饰，点到为止。

The second aerial view of the Technology Plaza. This picture emphasized the memorial landscape of Cherry Avenue specialyl designed in this plan and the relationship between public lawns and urban roads (pavement). The screen is without any decoration more than needed.

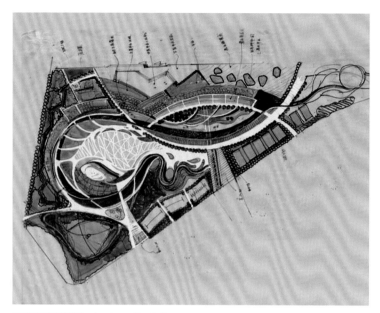 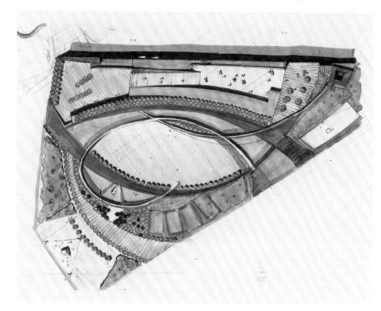

多层叠加的过程草图 Overlapping working drafts

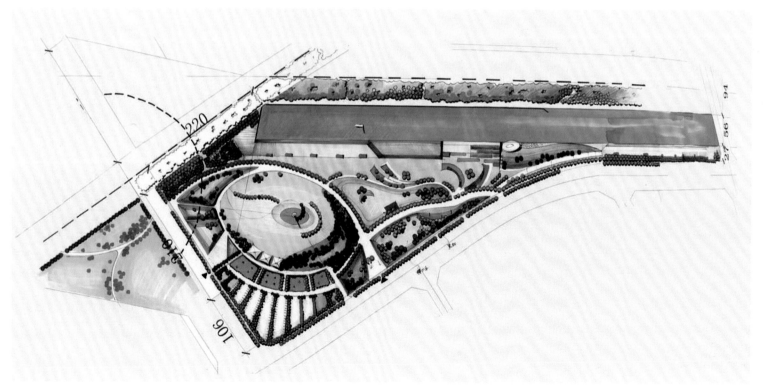

注释：
场地的东侧和南侧临曹妃甸新区污水处理厂和迁曹铁路高架线，污染、噪声严重，三轮设计均以地形和林带的遮挡为主，另外两面面向城市开放，同时将公园入口偏离城市交通岔口，在两入口之间设置内向停车场。

The Caofeidian wastewater treatment plant and elevated railways of Qiancao Line were adjacent to the east and south side of the site respectively. They made the site polluted with noise and contaminations. All three rounds of design focused on the shelter function by forest and terrain and left the other two sides open to the city. The entrance of park was deviated away from urban crossroads. An inward parking lot was located between two entrances.

总平面 Master plan

以色列耶路撒冷城市景观设计国际竞赛
International Urban Landscape Design Competition in Jerusalem, Israel

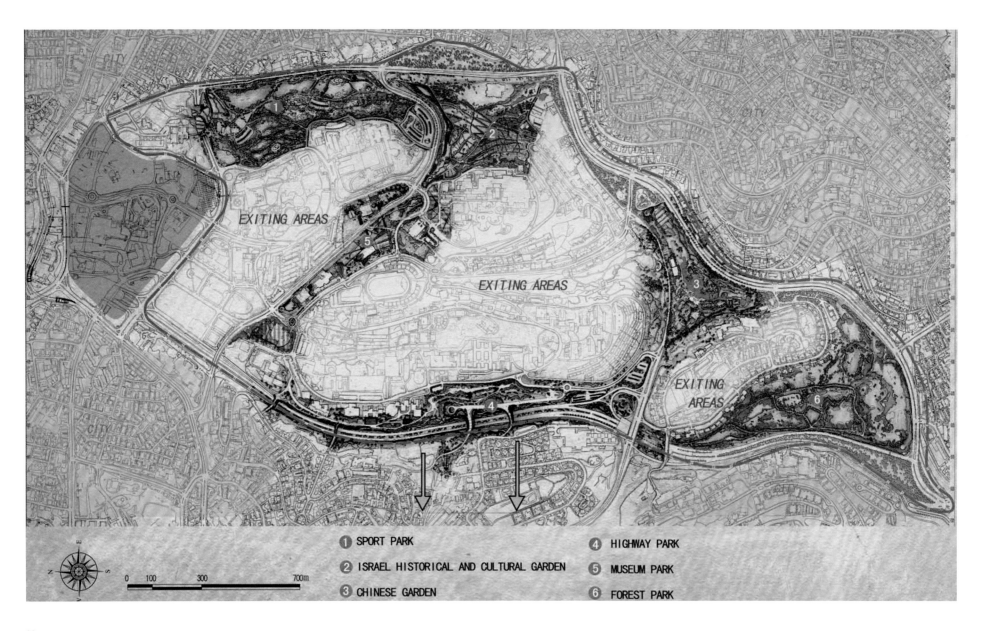

项目简介

以色列耶路撒冷城市景观设计属于"耶路撒冷城市之门"城市设计国际咨询的第二部分。项目所涉六块主要场地总面积约380hm²,位于耶路撒冷新城南部,环希伯来大学一带,场地毗邻以色列博物馆,死海古卷博物馆以及历经千年的中世纪教堂,是耶路撒冷旧城西部重要的文化古迹聚集区和高度敏感区域。场地地形由北部的台地到南部山谷,高差约150m,包含了山地、多层台地、山谷等天然景观。极为复杂的地形条件造就了形态丰富的景观基质,同时也给地形、水系以及交通等方面的设计带来挑战。

本设计依据现状地形的总体趋势,在380hm²的巨大场地上,由北至南布局了体育公园、博物馆园、历史文化园、高速公路园、中国园和森林公园等六个主体场地,分别容纳了城市休闲、运动、文化交流、教育等功能。

Project Summary

This urban landscape design in Jerusalem, Israel is the second part of the international consultation of "City gate of Jerusalem". The six main sites of about 380 hectares involved in this project, are located on the south part of new city region of Jerusalem, surrounding the University of Hebrew. Be in one of the most important historical relic accumulated area with high level of sensitivity in the west Old City of Jerusalem, the site adjoins the Israel Museum, Museum of the Dead Sea Scrolls and medieval churches which have gone through centuries. The height difference of the site from the north terrace to the south valley is about 150 meters. There is a variety of topographic forms including natural landscape like mountains, multi-platforms and valleys. The extremely complex geography condition provides a rich basement with diverse forms for landscaping but also a design challenge for site planning, water system, traffic and other aspects.

This design is based on the overall changing trend in terrain. Six main functional blocks were arranged on the giant 380-hectare site, which are Sports Park, Museum Park, History and Culture Park, Highway Park, China Park and the Forest Park, fulfilling the urban functions of leisure, sports, cultural communication, and education respectively.

设计思想

将场地作为以色列传统农业文化和地理版图的双重表征。

1. 充分尊重原有地形,并赋予场地地貌以独特的文化内涵,以景观的手法表达出以色列国土和地形特征。从最南端的森林公园到北端的体育公园包含了山地、台地、峡谷等复杂的地形变化,这种由高到低,由山到水的变化,恰似以色列国家版图的微缩,代表了以色列国土的基本特征:由高山圣城耶路撒冷,穿越绿色的山谷和平原,奔向地中海,直达沿海低地特拉维夫。由北向南,景观序列中的一座座花园、公园和博物馆犹如朝向圣城景观之旅中的节点,将场地中现有的文化遗迹、历史信息结合为一个有机整体,而朝圣之旅中的每一次停驻,都是历史与现实的交汇和重新诠释。

2. 植物元素的文化意义:植物设计以以色列特有的石松、柏树、橄榄树、柑橘、棕榈、葡萄、薰衣草、无花果为主导,突出以色列悠久的农业文明和发达的现代农业科技。在六个主要花园中不断重复出现的特色植物是对场地的呼应,也是对以色列悠久农业文明的颂扬。

Design Concept

The site is endowed with the dual representation of Israeli traditional agriculture culture and geography of the territory.

1. Fully respect existing local topography and endow the site terrain with unique culture context by expressing Israeli territorial features in the language of landscape. There is a variety of terrain types from the Forest Park in the north side to the Sports Park in south including mountains, multi-platforms and valleys. This high-to-low and mountain-to-water change is like the scaled model of the whole nation territory and represents its characteristics: begins from the holy city Jerusalem on a mountain, runs through green valleys and plain, heads to the Mediterranean and finally reaches coastal city Tel Aviv on lowland. Gardens, parks and museums located on the landscape axis from north to south, just like the serious of scenic spots on a pilgrimage to the holy city. Existing cultural remains and historic information was intergraded as an organism while every stop on the pilgrimage is being the intersection and re-explanation of history and reality.

2. Cultural significance of planting features: design about planting were based on special local spices like Lycopodium, cypress, olive tree, tangerine, palm, grapes, lavender and fig-led. It stresses the long-history agricultural civilization of Israel and the advanced modern agricultural technique deliberately. The repeating feature plants in the six main gardens are the correspondences with the site and the hymn to Israeli age-old agricultural culture.

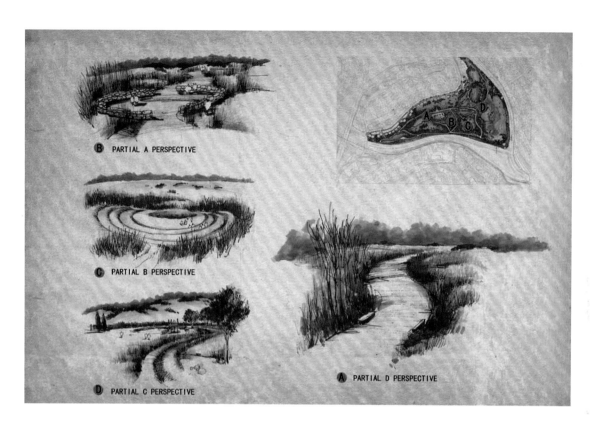

B PARTIAL A PERSPECTIVE

C PARTIAL B PERSPECTIVE

D PARTIAL C PERSPECTIVE

A PARTIAL D PERSPECTIVE

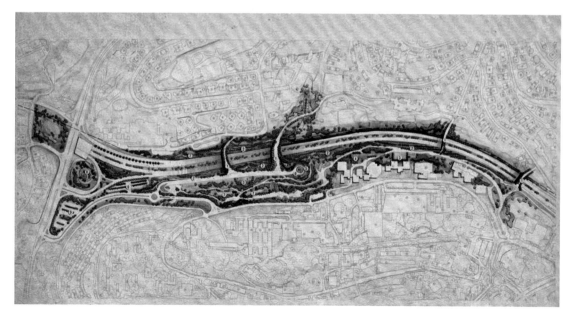

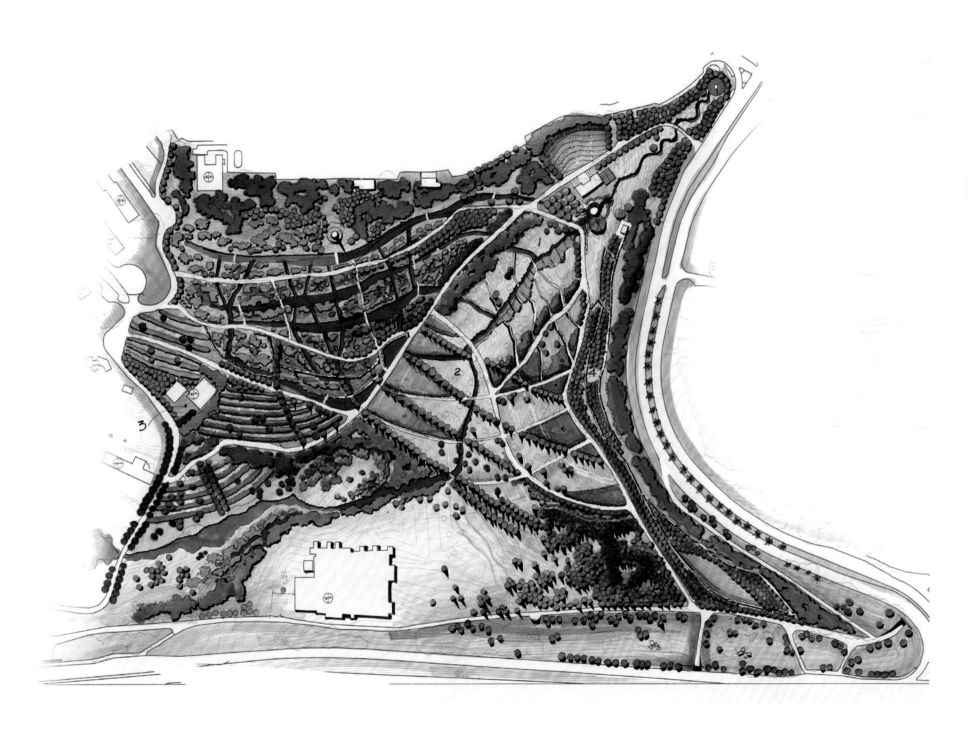

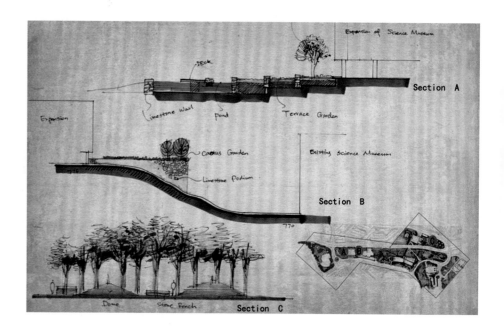

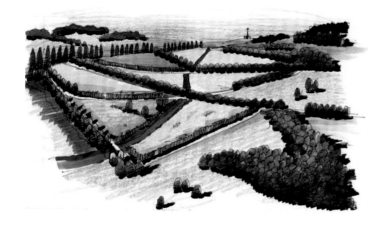

以色列历史文化园

　　历史文化园植物设计的灵感来自于圣经中提到被认为是富有以色列地域性的七种水果和谷物，包括小麦、大麦、葡萄、无花果、石榴、橄榄和椰枣。除此七种植物，在保留现有植被和地形的基础上，还加入了其他一些当地树种，象征了以色列的农业、历史和文化传统。通过符号和象征传达出场地的历史感和文化上的延续性，在保护历史遗迹和现状植被的前提下，营造出适宜人们享受大自然并能缅怀过去的平和欢乐的花园。

　　用水渠划分场地、联络空间的做法充分考虑了场地的历史感和文化属性，考虑到耶路撒冷这座对于犹太教、伊斯兰教和基督教均具有重要意义的城市所应有的景观形式和能反映场地历史文脉延续性的文化符号。与东方的中国园林和日本园林所追求的不规则布局、自由形态相反，伊斯兰园林和基督教文化下的园林基于一种精心设计的直线型和四分园的基本构成模式：由中心引出

Israeli Historical and Cultural Park

The inspiration of Historical and Cultural Park are the seven kinds of unique Israeli fruit and grains mentioned in the Bible, including wheat, barley, grapes, figs, pomegranates, olives and dates. Some other local trees were also added in the landscape besides these seven species, on the premise of protecting existing vegetation and terrain and presenting the agriculture, history and traditions of the country. We are expressing the sense of history of the site and the continuity in culture by symbols and signatures. With premise of protection on historic remains and existing vegetation, a peaceful and joyful garden was created, where people can enjoy the nature and cherish the memory of the past.

The site is divided and linked by drains. This practice takes full account of the history and context of the site. The drain is regarded as the perfect

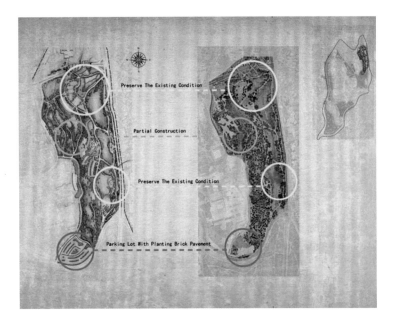

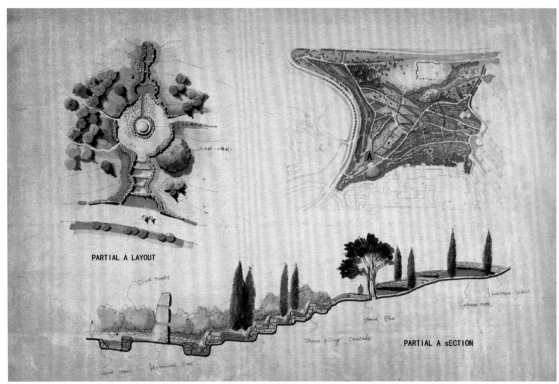

的四条水渠伸向花园四方，将一块场地划分为四个方形园林或苗圃，以此象征天国世界的基本模式。

这四条水渠在印度莫沃尔王朝的伊斯兰园林中，被称为"水、酒、乳、蜜"四河，而四条"河流"交汇之处必然会有甘洌的清泉喷涌而出，这便是人类的天堂。这样的园林景象与中东地区骄阳似火，漫天尘沙的自然环境形成鲜明对比，恰恰反映了在极为恶劣的自然环境中的人们对天堂的想象。水渠划分的四分园模式本身，对于饱受瘟疫肆虐的欧洲民众而言，就象征着避难的天堂。而今人在设计、建筑和养护这样一座园林的同时，相当于在地球上打造出了一处小小的"天国"。

symbol and landscape form to represent the historic continuity of the site in Jerusalem, a holy city with significant meaning to all Judaism, Islam and Christianity. The pursuit of eastern Chinese and Japanese gardens is irregular layout and free-form. Totally different with that, landscape in Islamic and Christian culture rooted from a delicate linear and quad-division structure: four drains originated from center head to the four sides of garden and divide the garden into four square gardens or nurseries, which is a symbol of the basic form of heaven.

These four drains in the Islamic garden of Indian Mughal Dynasty were known as "water, wine, milk and honey". Wherever these four rivers meet, there will be fresh spring spouting out whcih turns to be the paradise. This landscape scene is a strong contrast against the natural environment of the sweltering heat and flying dust in the Middle East region and expresses the imagination for heaven from the people who live in an extremely harsh environment. The garden mode which is divided by drains into four parts is a symbol of sheltered paradise for the European people suffering form the plague. When you design, build and maintain such a garden, your efforts are just like to create a little heaven on earth.

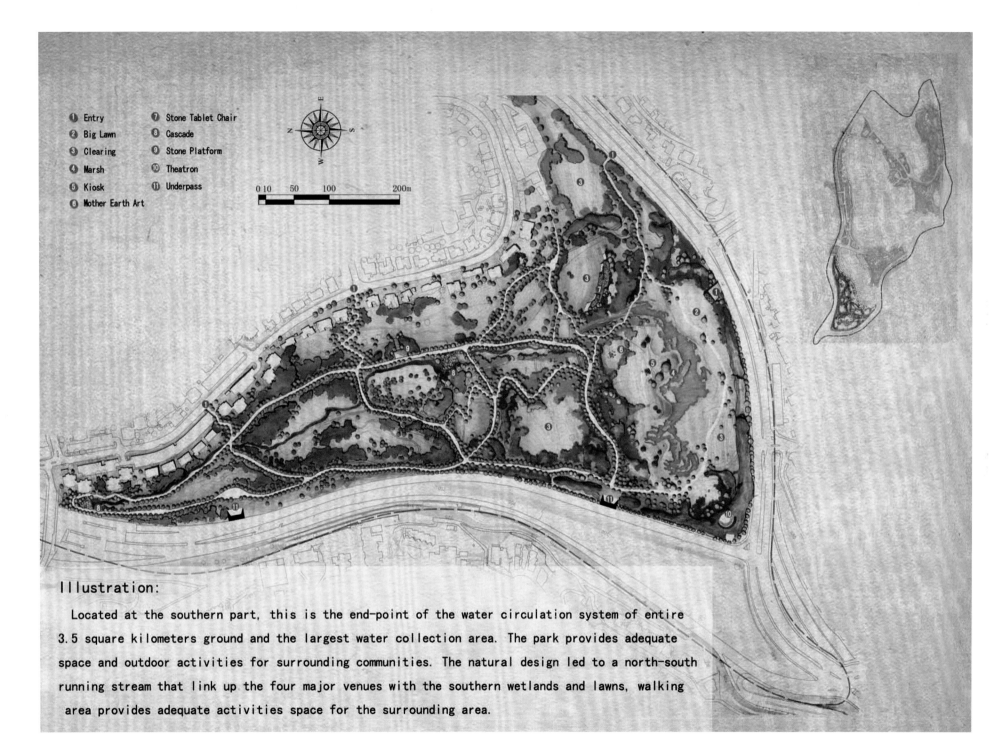

Illustration:

Located at the southern part, this is the end-point of the water circulation system of entire 3.5 square kilometers ground and the largest water collection area. The park provides adequate space and outdoor activities for surrounding communities. The natural design led to a north-south running stream that link up the four major venues with the southern wetlands and lawns, walking area provides adequate activities space for the surrounding area.

■ 中国园

China Park

中国园建在一片四周高中间低洼的三角地块上,自中央汇水面设主湖,并使之成为园中所有建筑和假山叠石的共同背景。围绕主湖和向西、南伸展出的两条溪流、叠水串连了三组庭园和建筑,即大唐中国园、苏州园和一个典型的北京四合院庭院,分别体现轴向、非轴向自由空间和围墙为特色的三种典型的中国容积空间形式。三组空间成为展示多种历史悠久的中国工艺、文明和民俗的舞台。具有中国园林特色的植物,如黑松、白皮松、侧柏、牡丹、竹、梅和最重要的石组等,将成为景观空间最主要的表情,将原植物园的植物基底延伸入三座庭院外围,形成新旧景观之间的联系和桥梁。这座东方风格的园林最终将融入植物园景观之中,并以自身特有的方式与原有的风景取得一种微妙的平衡。

China Park was built in the middle of a bowel-shape delta. An accent lake was designed to gather all water at the central bottom to be the common background for all the buildings and landscapes in the site. Two streams rise from this accent lake head to south and west. Three groups of gardens and buildings are linked by the two streams, which are Tang Dynasty Chinese Park, Suzhou Garden and a typical Beijing quadrangle courtyard. They represent three kinds of Chinese typical spatial forms: the axial space, irregular free-shape space and enclosure space, respectively. They are the platform to exhibit the long-history Chinese arts, crafts, culture and traditions. Plants with unique features of Chinese landscape will be the main characters in these scenes, such as pine, bungeanas, arborvitaes, peonies, bamboos, plums, and the most important rock groups. Existing vegetation of the site will extend into the outside of these three gardens to connect old and new scenes. This oriental garden will eventually integrate into the landscape of the botanical garden while getting the balance with existing scenes in its own unique way.

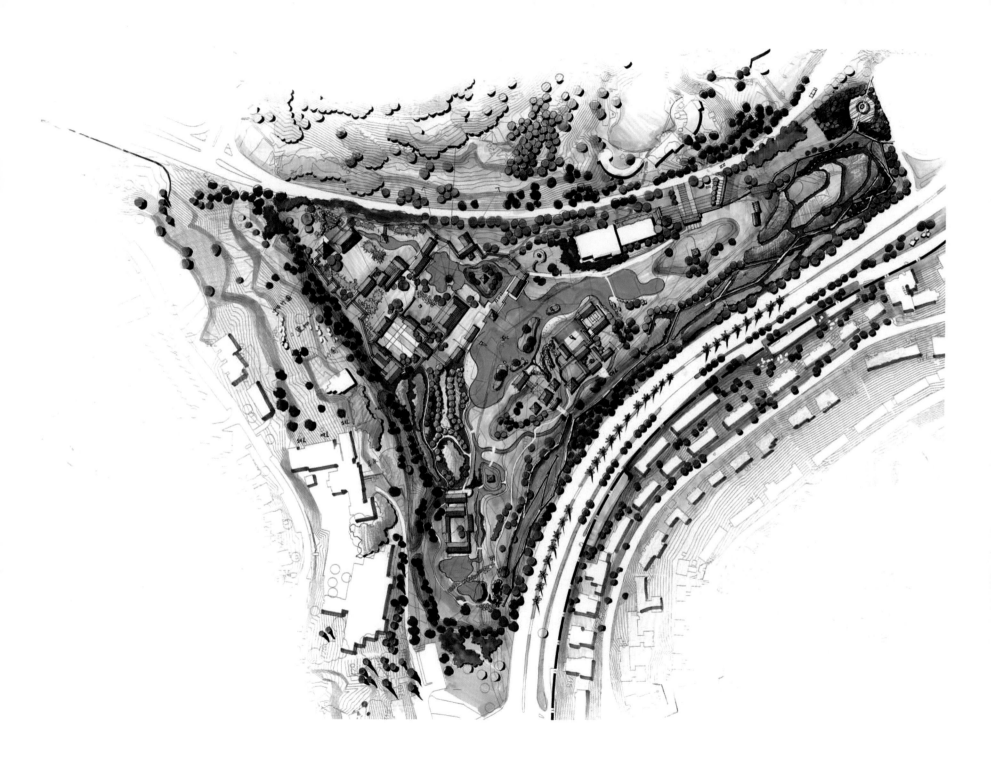

米巅宋梓龍石箑浦岳，昔廬東長岳壽山有壽石曰「」靈璧、林屋、昱壁、美壽。為廬。見為狀也。嶽口之綠綺。蜀僧抱琴之苦思君所之江浙。鲫舫千里，馬蹄相逢，力身天遠，皆動之江南。此四「同之衔塍綱媽春摩恣氣窑，真南蹌湾不僵具腔。宗城靜而之。史填尺。一任為武橋官室，奥应圃山立官之張棠冲，御雲萬魔，除秀才怪江人嵌岚美。或。「青慶文間十一月大姜陪，郡人指嚳嵴楔適房于壽山長嵌之」畢蠻袒秀。蘆陽宗岌。靈景之美，見秋之春信唐之垣目。長男石湾美人箝江人爲故美古林亀。陵差萬年。偽天下之隙觀。古今之勝在焉……西武箴新春。富青之参貫圃長。宜加酒四。自后。食得三玠石。後乎江南。长岳為遠。何見。

地之浦，仁為年波樹海浦，鐘
景。嚴宣之姿，嘉頭誘天恋。
力可為。唐人摆之山。
以珊瑚。宋人姓之
將名為兄。為文
文烟女士姐之秋
反政。尤是也含神
為烏。剔遊自瓷缾
二曁山。善隐化之
云。剔進凰目宗代心余
園園憂達風何此為
之伊靡。由一石可見矣。
憂之奈。奥遑餘止。圃人
長海矣。墨箆之人随粤

石譬驗宋梓龍石箑浦，奏餒
昌繇唐史坎石，意為遠筆草
渐五忌茅年犁，為名河沈笑貸，
不藏蓮拷，歟神，涓月柔于爲～
苕春擧畫，昕首宲戯之志摧在，
芭天生淌体音春拭手工藝，碎之入堂
奥，藏者礌也，函狮傅心文哉。

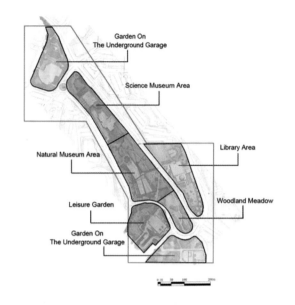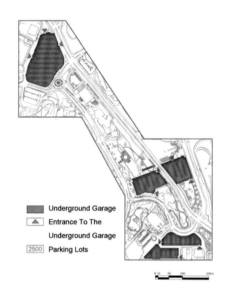

■ 博物馆园

■ Museum Park

中心博物馆区，包括博物馆区、科技馆、自然博物馆、图书馆、死海经卷博物馆和科技馆扩建范围，以及外围支撑服务体（两块绿地，多个小规模点状喷泉和半圆、三角形水池）组成。博物馆园提供三个大型展馆区的集中人流疏散和休闲绿地系统，由多条线状水渠和地下管道串连，并与完善的步行系统相结合。所有入口、停车场及地下停车端口均与步行道系统构成完善的联络——线状系统。由喷泉、柱廊、绿丘、雕塑装饰的多个景观节点，主要包括：两座与水渠相连的喷泉，一座爱奥尼柱式的半圆剧场，一个雕塑花园、仙人掌园、平台花园和三座由石灰岩围合、分割的台地花园。使用当地盛产的石灰岩为主要材料，构筑花台和多层次的地台，配合规则排列的植物，将这一西方园林的传统以具体的形式表达出来。运用富于中东和地中海特色的橄榄、棕榈、石松、柏树为主要遮阴树。浓郁的石松、侧柏以其独一无二的剪影提示了以色列文化与欧洲文明以及地中海文明悠久的联系性。

The central museum zone consists of the Museum District, Science and Technology Museum, Natural History Museum, Library, Museum of the Dead Sea Scrolls and the expansion area for Science and Technology Museum, as well as the external support services (two green areas, a number of smaller point-like fountains and semicircular, triangular pools). Evacuation area and recreational green space for concentrated flow of people from three large exhibitions are provided and integrated with the comprehensive walking trail system by linear channels and underground paths. All the entrances, parking areas and the entrances to underground parking area are connected to walking trail system properly to form a linear system. Scenic spots decorated by fountains, colonnades, green hills and ornamental sculptures include: two fountains connected to the channel, an Ionic semicircle amphitheatre, a sculpture garden, a cactus garden, roof garden and three terrace gardens divided and enclosed by limestone. The traditional Western landscape features were explained in specific forms, such as the use of local rich limestone as the main material to build flower nurseries and multi-level terraces and to match up the regular arranged plantings. Main shading trees are olives, palms and Lycopodium which are iconic plantings in the Middle East and Mediterranean region. The thriving Lycopodium and cypress with its unique figure hints the long-history relationship between European civilization, Mediterranean civilization and Israeli culture.

Shuyu Cultural Park in Sanhe, Heibei Province
河北三河市漱玉文化公园

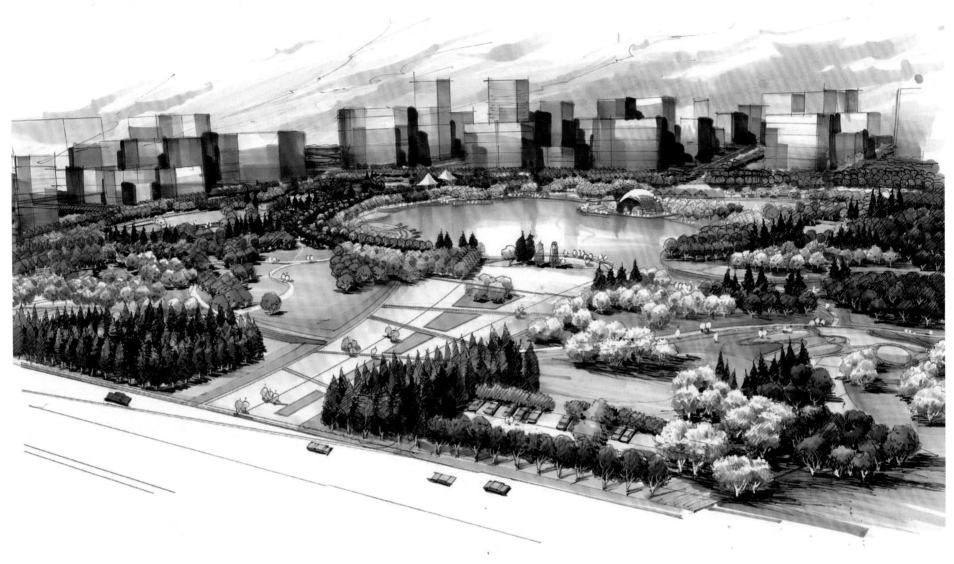

北入口鸟瞰图　Aerial view to north entrance

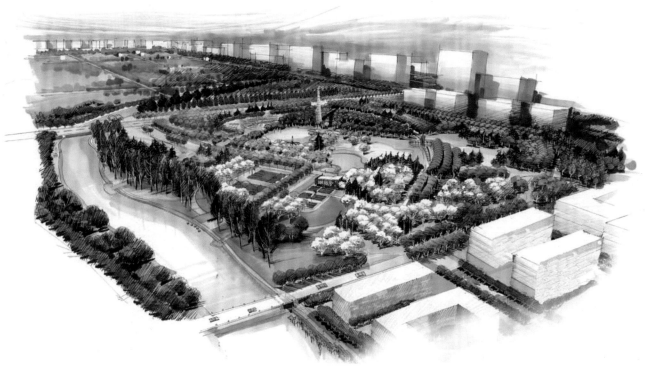

滨水区鸟瞰图　　Aerial view to waterfront zone

南入口鸟瞰图　　Aerial view to south entrance

北京台湖镇新城绿地景观规划
Landscape Design for the Green Land of the New Town of Taihu Town, Beijing

台湖新城绿轴中央区总平面　　Master plan for central zone of the green axis in the new town of Taihu

注释：
　　此方案为北京台湖新城中央公园绿地轴线的第二轮工作草案。核心思想是利用场地轴线北段地势太低的现状，形成连贯的南北向城市绿轴，并依据城市绿地规划划定的绿地范围，将规划公园与现状绿地结合，形成了未来新城完善的绿色网络体系和城市市民休闲公园体系。

This design is the second round draft for the central park green axis in the new town of Taihu, Beijing. The core idea was to improve the utility of existing north part of the axis in the site and form a run-through north-south urban green axis. The design is also based on the area requirement from urban plan; integrate the park in planning into existing green space; formed a perfect green network for the future new town and the park system for leisure and relax.

商业区意向　　Commercial zone imagery

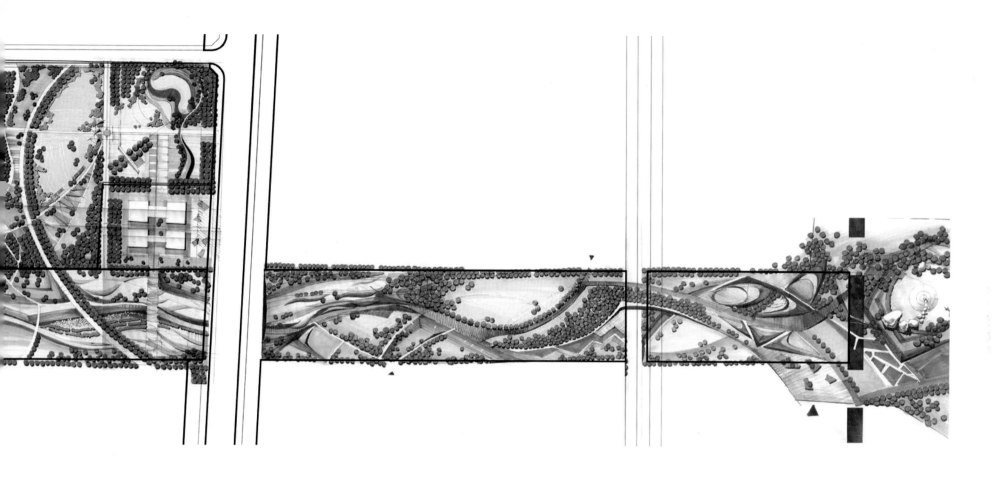

绿带途经各个新城街区及放大区域公园区域的鸟瞰意向　　Aerial view to the green axis passing through blocks in the new town and enlarged park zone

迁西滦河——滦县东岸规划
Planning for the East Bank of the Luanhe River, Luan County

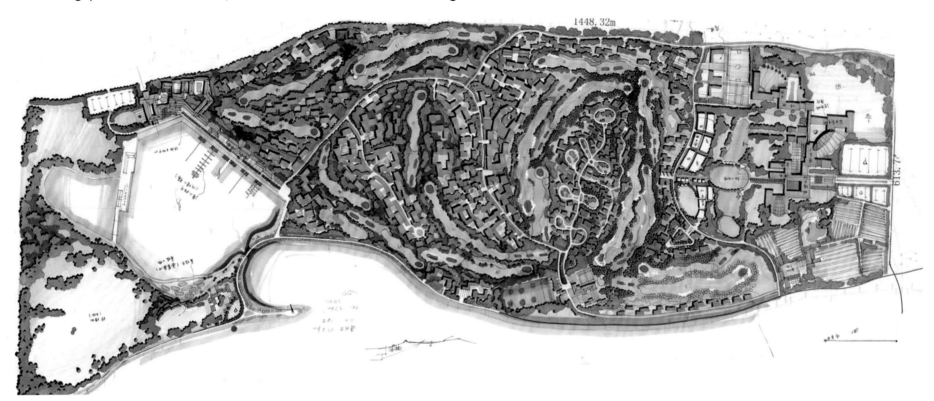

注释：
　　规划场地位于河北省滦县滦河东岸，北面隔着滦河与有千年历史的滦县历史文化地标——岩山塔相望。场地背山临水，视野开阔。在一百多公顷的开阔场地上，覆盖有大面积次生林和水湾池塘，自然条件十分优越。规划目标是在百余公顷的滨水岸线上，建设高标准的企业总部园区和生态示范区。

　　主要功能区包括：一座作为园区中心的度假小镇，内设宣传展示公司产品的大型室内外展场、中央生态公园、体育休闲公园、一座大型生态停车场和一座小型示范农庄；独栋和合院式别墅为主的企业高层人员集中住区，其中包括一座十八洞标准球场的高尔夫场地；一座面向公众休闲度假的游艇俱乐部等子项。

　　分区设计：北区由天然水湾改造，结合动感渔人码头休闲小镇建设，形成北区中心；中区为高尔夫别墅，采用多向尽端路布置单体和院落式别墅，沿球道隔离展开，并结合人工微地形设计，形成高尔夫别墅的基本构成框架，组团道路沿东西向展开，为每一栋别墅争取优越的日照条件，使每一栋别墅都能看见球道；南区为企业综合展示区，设高科技生态农场和集团产品展示区，围绕现状湖面形成南区中心，入口设大型集中停车场。

The site in plan located on the east bank of the Luan River in Luan county, Hebei Province. A millennium-old local cultural landmark—Yanshan Tower—can be seen from the site crossing Luan River on the north. This site sits is against the mountain and faces the water with a great spacious view. This 100-hectares open space is covered with large area of secondary forest and ponds, being in excellent natural environment. The goal of planning is to build high-standard corporate headquarter zones and ecological demonstration zones along this hundred-hectare waterfront bank line.

Major functions includes: a resort town as the center of the park, which consists of a large-scale outside and indoor exhibition space for the products of the company; central ecological park, sports and leisure park, a large ecological parking lot and a small ecological demonstration farm; residential district for senior corporate staff, mainly single family villas and common-yard villas; a standard 18-hole golf course in the residential district; a yacht club for public leisure and entertainment ,etc.

Design of subzones: The north zone was transformed from a natural bay. The construction of the center of north zone was combined with leisure town of dynamic Fisherman's Wharf. The middle is a golf villa zone, where the single family villas and common-yard villas were arranged along single-end roads. The houses and villas were aligned along the isolated golf course and integrated with the artificial micro-topography design to form the basic structure of this golf villa zone. The district road extends to east and west. Each of the villas and houses along the road was provided with optimized sunshine and a view to the golf course. The south zone is an integrated exhibition area for enterprises, including high-tech eco-farm and exhibition for products of the Group. The existing lake was surrounded to form the center of south zone. A large parking lot was provided near the entrance.

由滦河看东岸的高尔夫别墅区　　View from Luan River to the golf villa zone on the east bank

从北侧游艇码头看规划场地，远端为滦水岩山景区地标——岩山塔

Perspective from the yacht club in the north to the planning site, Yanshan Tower—landmark of Luan River Yan Mountain scenic area in distance

高尔夫别墅效果图　　Image of the golf villa

唐山雕塑公园
Tangshan Sculpture Park

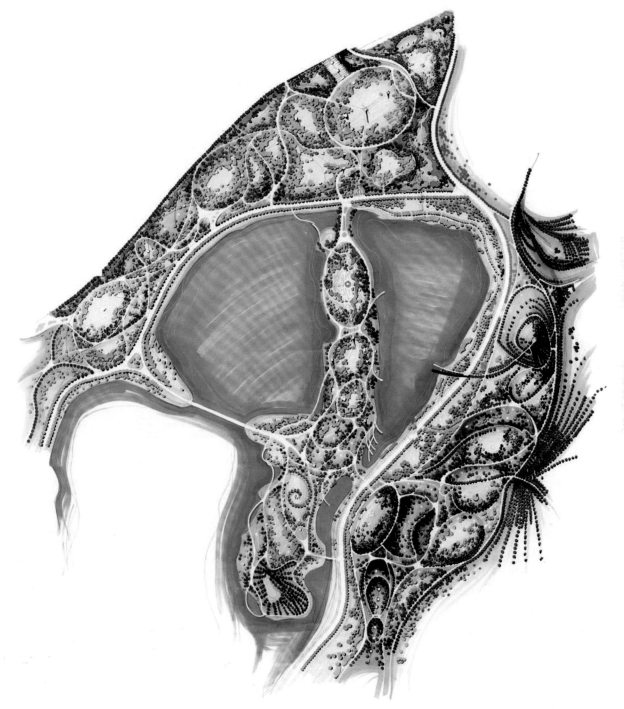

注释：
　　唐山南湖在第一阶段生态重灾区环境治理基本完成的基础上，进行产业结构重新布局的重要步骤，通过一系列国际型展会引入，再次提升该地区的综合环境质量，规划强调在对原有景观空间充分利用的基础上，针对展会要求完善景观层次，增加游览服务设施，并为雕塑艺术品提供适宜的展示背景和场地空间。

Tangshan government treated the ecological disaster in South Lke District in the first phase. Then the industrial structure was reconstructed by the introduction of a serious of international exhibitions. This is an essential step for the rearrangement of the industrial structure, through which the overall environmental quality was improved further. The planning stresses the utility of existing landscape and then rich the layers of landscape according to the requirement of exhibitions. Service facilities were set up and appropriate display space was provided for sculptures.

唐山南湖雕塑公园规划平面　　Planning layout of the South Lake Sculpture Park in Tangshan

注释：
　　樱花作为主景，常绿松柏为背景形成色彩上强对比对应和场所的氛围。樱花开放期极短，盛放之时绚丽繁华之极，但凋零时则转瞬即逝。恰如对故去者的回忆和哀思，也是对人生繁华过尽，最终归于平静的一种隐喻。

The background of evergreen pines and cypresses forms a strong contrast against the main scene of cherry trees to stress the space atmosphere. Cherry blossoms have a very short but splendid period and fade and consume away quickly and sadly, just as memories and grief to the deceased people. It is also a metaphor of the final calm life obtained after bustling.

雕塑公园凤凰涅槃纪念园意向　　Phoenix Reborn Memorial imagery in a sculpture park

林下休闲空间效果　　Effect of the leisure space in forest

注释：
　　凤凰纪念园西向南湖的入口，突出了景石引导的视景线。观者的视线穿越南湖，直达市民中央广场标志性的巨型凤凰雕塑。

West entrance to South Lake of Phoenix Memorial Park
This picture highlights the visual guide line of landscape stones. The viewer's attention is guided through the South Lake, direct access to giant sculpture of Phoenix landmark on the public Central Plaza.

凤凰纪念园多角度小鸟瞰　　Bird's view from different angles of Phoenix Memorial Park

南湖中央广场雕塑　　Sculpture on South Lake Central Plaza

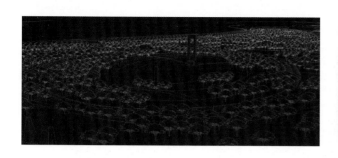

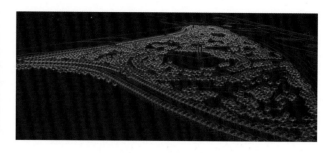

雕塑公园东北区世界雕塑公园展预定场地鸟瞰　　Bird's view to Northeast Exhibition site in Sculpture Park

秦皇岛东郊公园
Eastern Suburb Park in Qinhuangdao

南入口鸟瞰　　Aerial view to south entrance

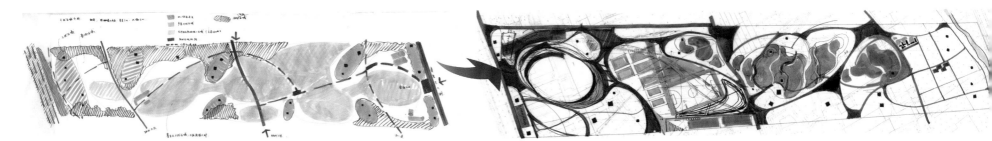

功能分区图　　First draft: Function zoning diagram

农业休闲区鸟瞰　　Aerial view to the agriculture recreation zone

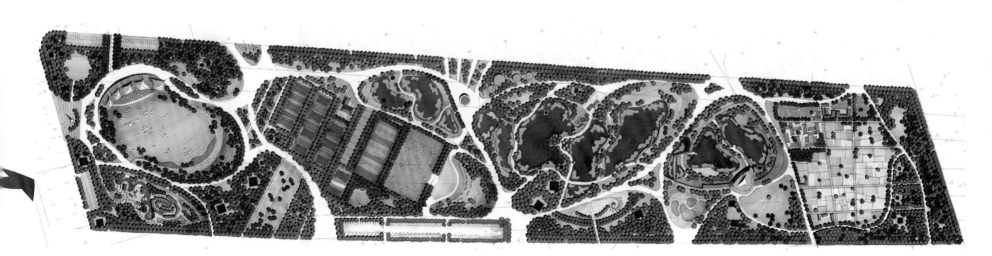

总平面图　　Master plan

秦皇岛某培训中心
A Training Center in Qinhuangdao

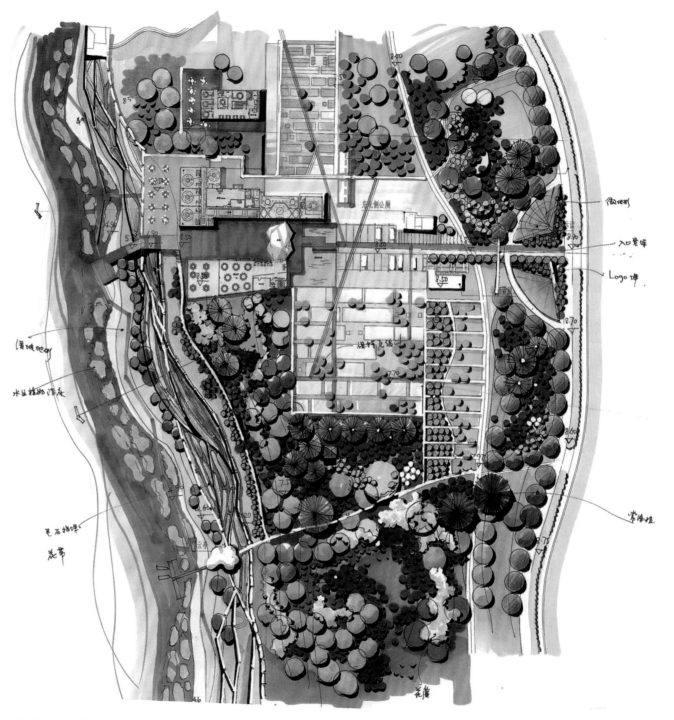

注释：

此设计为景观改造提升项目。场地河沿岸展开，竖向高差达5m以上。临河一侧多为陡坡，坡岸植被良好，并有适合设计湿地观赏花园的大面积滩涂地，丰水期部分可被淹没。

设计内容包括：对原有硬质苗圃式植槽加以改造，加入南北向连接道路，并在道路一侧增设停车场地。

利用周边现状植株，增设林下植物层次，形成宾馆周边的休闲林区，适应夏季旅游休闲的需要。

在沿河陡坡的一侧，通过地形树立，整建多层次观赏地花园，花带之间串联形成最佳沿河风光观赏区，下设水花园，成为场地重要的亲水活动区域。

设计特点：竖向上的丰富变化和场地适宜性设计。

This design is a landscape renewing and upgrading project. The site lies along the river with a height difference up to more than 5 meters. Part of the site along the river has steep sloping and good vegetations on it. It is suitable for an ornamental wetland garden and a large area can be partly submerged during wet season.

Design content includes: Renew the original hard-surface nursery planting tank, add north-south roads for connection and provide parking area along the road;

Add a lower layer of vegetation to the existing botanic environment; create a leisure forest around the hotel to meet the needs of tourism in summer;

On the sloping side, build multi-layer oriental garden according to terrain condition; link the flower belts together to form the best bank side scenic zone; a water garden is set at the bottom as an important water activity zone for the site.

Design features: versatile in vertical and suitability design for the site.

总平面图　Master plan

沿河的观赏水花园及陡坡上的台地花园　　Ornamental water garden on waterfront and terrace garden on a steep sloping field

秦皇岛洋洋花海
Yangyang Flower Field in Qinhuangdao

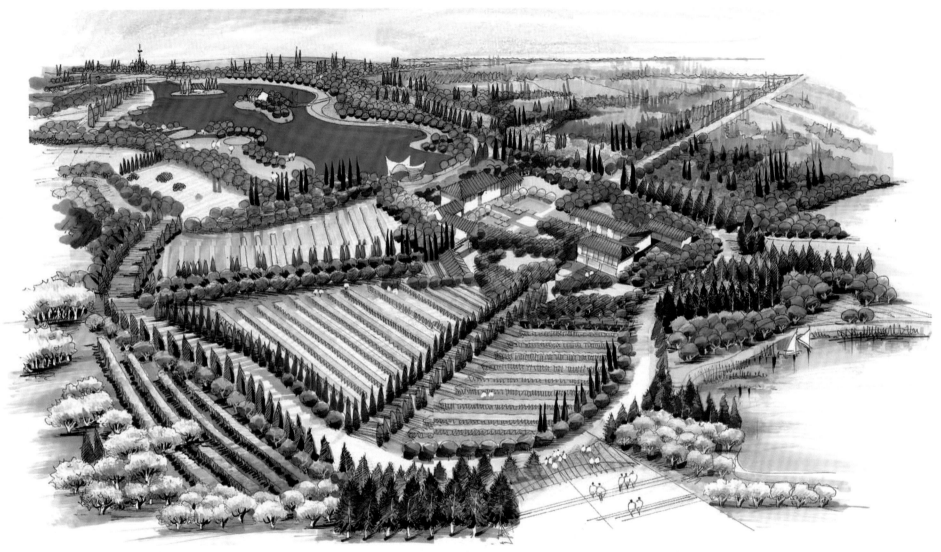

花海鸟瞰，近端为酒庄及葡萄酒种植景观意向，远端为农场主湖区 Aerial view to the flower farm. The nearby is the landscape image of grape plantation and the vineyard. Main lake area in the farm appears in the distant view

葡萄酒庄园入口主轴透视　　Perspective to the entrance axis of vineyard

酒庄葡萄园　　The chateau of the grape plantation

酒庄葡萄园　　The chateau of the grape plantation

辽宁铁岭新城凤冠山景观设计

Landscape Planning for Fengguan Mountain in the New Town of Tieling, Liaoning Province

注释：

标志性景观区域的设计与表达，场地位于山顶，可鸟瞰主湖和周边新城，视野开阔，不仅是场地轴线的端景，也是重要的观景区。"看"与"被看"的特征非常明显。

设计采用传统中国园林的景石和地方特色植物——黑松作为背景，形成具有标志性和观赏性的集中展示区。

The design and expression for the landmark scene. This area located at the top of a hill with an open view to the main lake and surrounding new town. It is not only the ending scene of the whole site's axis, but a spot for see-sighting. It is a typical place "to see" and "to be seen" at the same time.

This iconic and ornamental concentrated exhibition zone is decorated by traditional Chinese landscape garden feature stones and the unique local plant-black pine as the background.

唐山南湖——唐胥路

Tangxu Road, Nanhu in Tangshan

市民广场—中央大草坪区、滨湖演艺区　Public Plaza-Central meadow and lakeside entertainment zone

注释：

唐胥路为唐山南湖景区中唯一纵身超过百米的道路，28km² 的景区中主要的功能场地均集中于道路北侧。各级领导高度重视这次项目，将之视为解决南湖风景区、停车、集散和市民交流的一系列计划中的决定性步骤，对该场地的详细设计从 2009 年开始，一直持续到 2011 年，共经过六轮方案修改。至今年，初步确定方案，此为设计过程中第二个方案草图，反映的共同特点是：保留北侧沿湖岸上面，尽量多争取沿湖场地以留给市民活动，并设置多条步行和环湖自行车的游览线路。并且，为未来几年陆续展开的世界雕塑园展和园林博览会留下一定的场地和服务体系。

Tangxu Road is the only road over one hundred meters in the 28km² Tangshan South Lake scenic. All the main functional sites in the scenic concentrated along this road. This project has gained the attention from all relative officials. It is regarded as the decisive solution to the problem of parking, gathering and public activities in South Lake scenic. The detailed design for the site was carried on from 2009 to 2011 through 6 rounds of modifications. Not until this year has the final plan been decided. This picture is the draft of the second design in this procedure, reflecting the common characteristics: to retain the north side of the lake, as much as possible in order to leave space for public activities along the lake. It sets a number of walking and bicycle tour routes around the lake, offering enough space and service system to the future global sculpture exhibition and garden fair in the next few years.

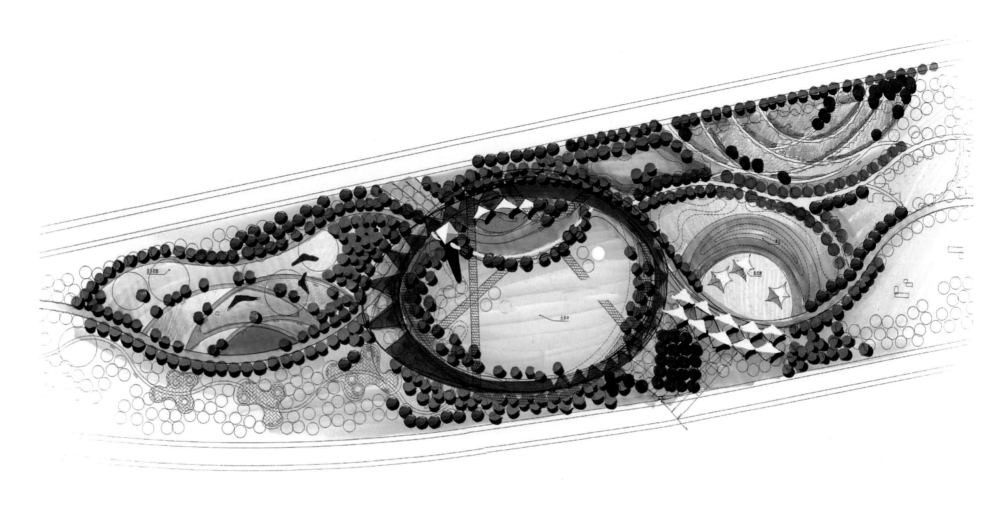

注重市民多种游乐休闲活动的方案之一　　One of the design draft that focusing on the diversity of public activities

■ 唐山南湖生态城——环城水系及新开河扩湖景观

Landscape of the Urban Hydrographic Net and the Enlarged Lake in Nanhu Eco-Town, Tangshan

注释：
　　同样的改造提升设计，目标是将原有的单一形态的硬脂浆砌驳岸重新设计，改造为适合于市民休闲和科普教育展示的滨水实地。实地的表达方式可以非常丰富、自由，相比于电脑彩屏，手绘图更为灵活、多样，在意向性植物表达方面仍有一定的优势。

The goal of this similar promotion design is to redesign the original single form of stearin mortar revetment into a waterfront suitable for public education, leisure and exhibition. Hand-drawing can be very versatile, free and rich, compared to computer rendering. It also has the advantages in expressing feature plants.

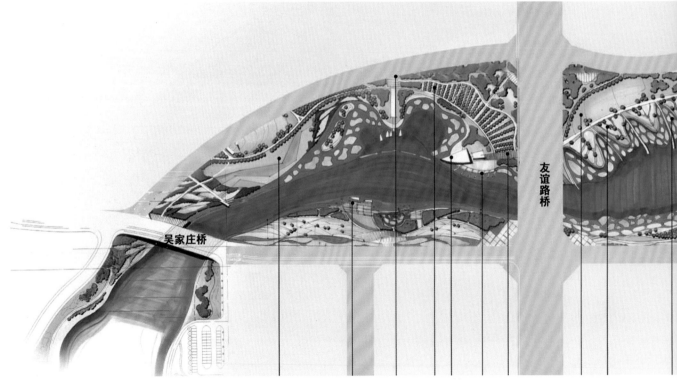

唐山环城水系新开河段扩湖项目总平面　　Mater plan of enlarged lake program of circling water network of Tangshan city

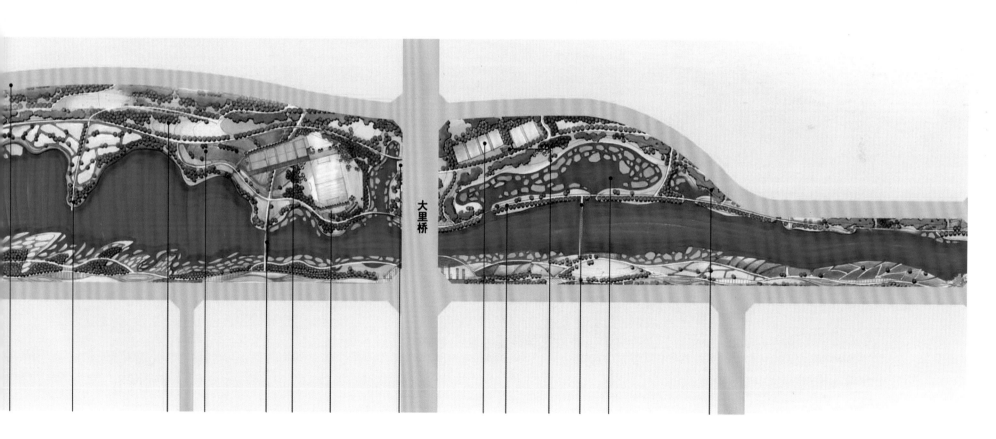

大里桥

西外环的迎宾路节点　　The left picture shows a node on Yingbin Road on the West Ring Road

唐山丰南引水渠入南湖段节点　　The right picture is a node where the Tangshan Fengnan division canal connects to Nanhu Lake

唐山南湖生态城规划
Urban Planning for Nanhu Eco-Town in Tangshan

注释：

南湖生态城第一轮总体景观规划及 6km² 起步区概念性城市设计总图。

总图中规划了：未来城市的总体钻石型结构和直达南湖中央的凤凰发展轴，形成严整的视觉对称和清晰的城市结构。

结合现有青龙河履带的城市绦轴并入唐山环城水系总绦网中。

初步规划了唐胥路以北的主要功能区，包括规划大南湖与现状小南湖公园的结合、植物园、唐胥路市民公园、青龙湖生态湿地，以及中央凤凰洲的标志性建筑区。

唐山南湖的严整轴线，对景设计以及主体绦网结构是在四年前规划第一份草图阶段就已经确定的，以后三年的规划更多的是对每一块场地的细化。可见创意性草图及工作模型对景观规划行业发展的价值。

The first draft of overall landscape planning in the ecology town of Nanhu and the conceptual urban planning layout of the zone with 6km².

In the layout, it shows the diamond-like structure of the city in future and the central axis of the Phoenix directly pointing to South Lake. A rigorous symmetry in vision and clear city structure is determined.

The existing city green belt of Qinglong River is integrated into the overall circling water network of Tangshan city.

Function of the area on the north of Tangxu Road has been planned primarily, including the combination of Big South Lake and the existing Small South Lake Park. Main construction area has been decided in Botanic Garden, Tangxu Road Public Park, Qinglong Lake wetlands and the central Phoenix Island.

The rigorous axis of South Lake in Tangshan and its role in landscape and main road network has been determined in the first draft since four years ago. The work in next following three years lays more emphasis on the refinement of each single site. This case shows the value of conceptual sketches and working models in landscape planning.

唐山南湖总体鸟瞰图　Overall bird's-eye perspective of Nanhu in Tangshan

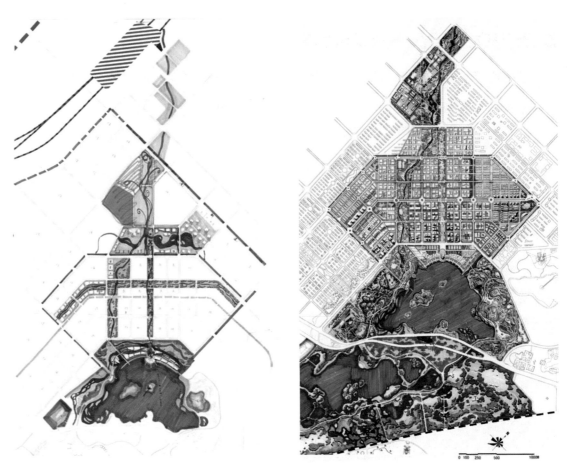

唐山南湖第一轮设计总平面　　The first draft of overall landscape planning in the ecology town of Nanhu

唐山青龙河城市绿轴第一轮设计总平面　　The first round master plan of the Green-Axis Design in Qinglong river, Tangshan

唐山南湖生态城——青龙河景观
Landscape of Qinglong River, Nanhu Eco-Town in Tangshan

驳岸效果图　　Waterfront Rendering

青龙河湿地及周边区城市设计　　Urban design of Qinglong River wetland and surrounding region

唐山南湖生态城——市政广场
Municipal Plaza in Nanhu Eco-Town, Tangshan

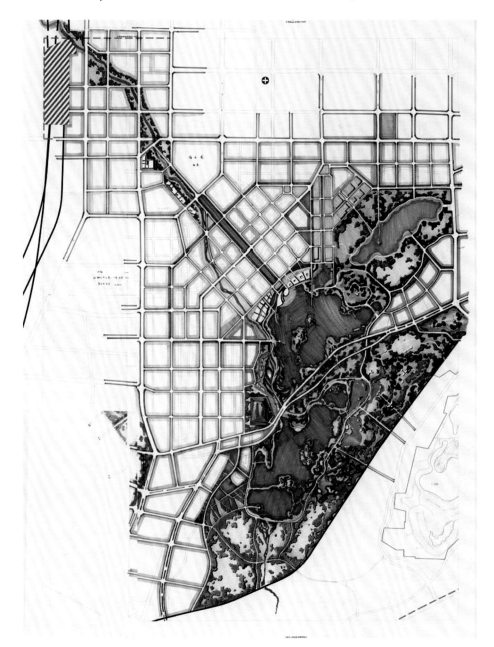

一草：南湖景观设计　　First draft: Landscape design for Nanhu

北区植物园规划　　Planning for the Botanic Garden in the north zone

南湖起步区城市设计　　Urban design for the entrance zone of Nanhu

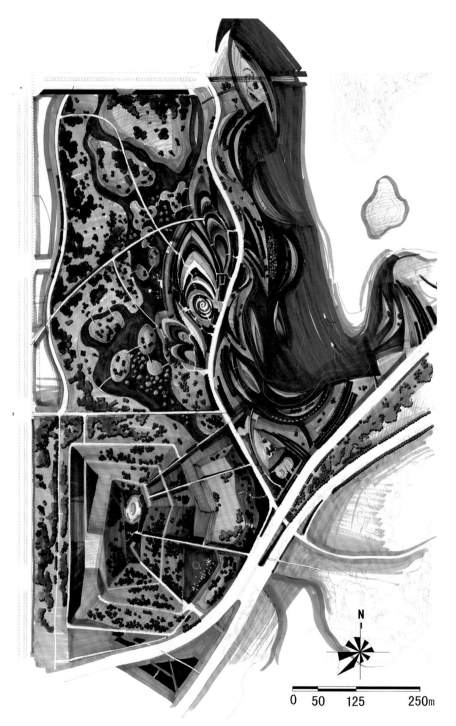

湿地科普区设计　　Wetland education area design

湿地科普区意向　　Intention of the science popularization area on the Wetland

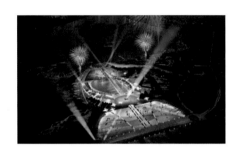

滨水草阶舞台　　The waterfront grass-cover amphitheatre

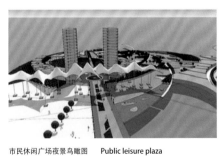

市民休闲广场夜景鸟瞰图　　Public leisure plaza

生态驳岸意向　　Imagery of the Ecological revetment

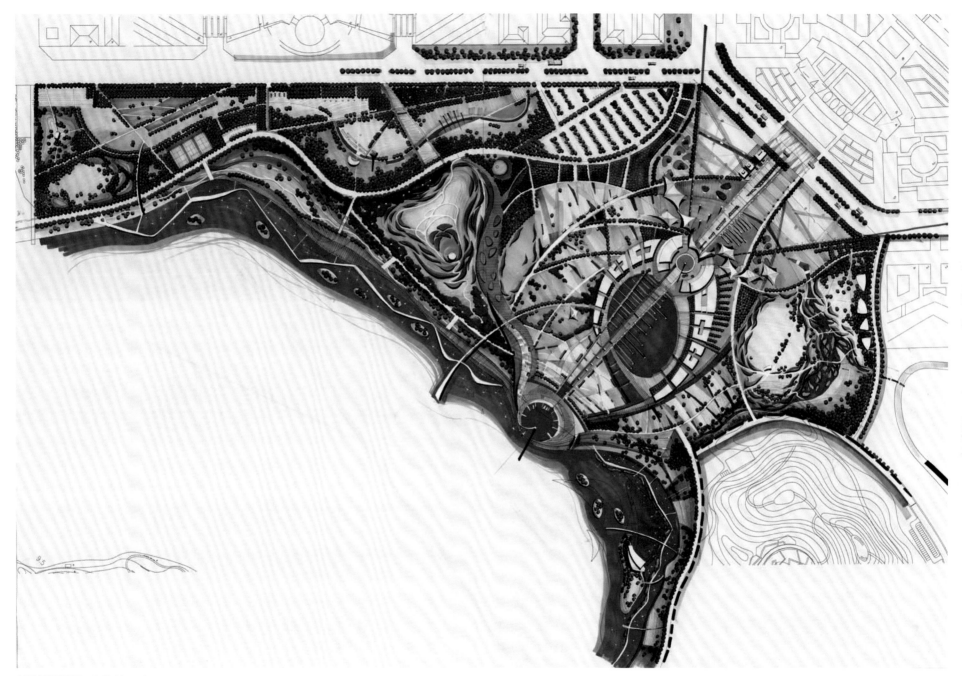

市民休闲广场总平面　　Public leisure plaza

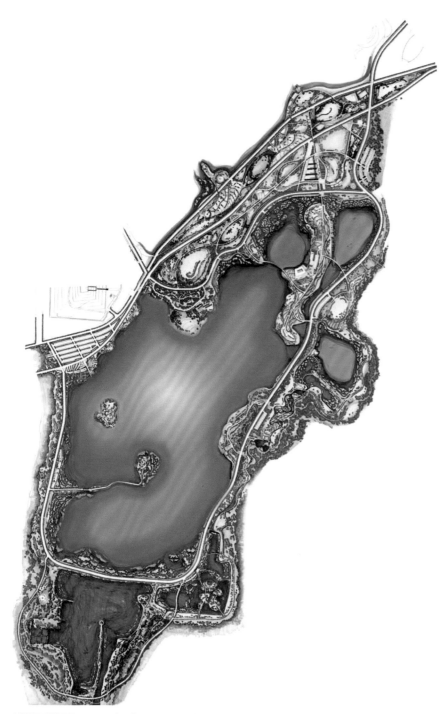

南湖南面区域总平面　Master plan of the south region of South Lake

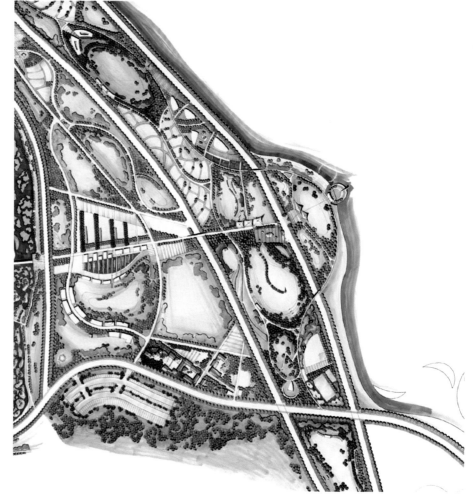

南湖中部世界雕塑公园展平面　Layout of global sculpture park in the middle part of Nanhu district

湿地科普区设计　　Wetland education area design

唐山植物园规划
Planning for the Tangshan Botanic Garden

唐山植物园总平面　　Master plan of the Tangshan Botanic Garden

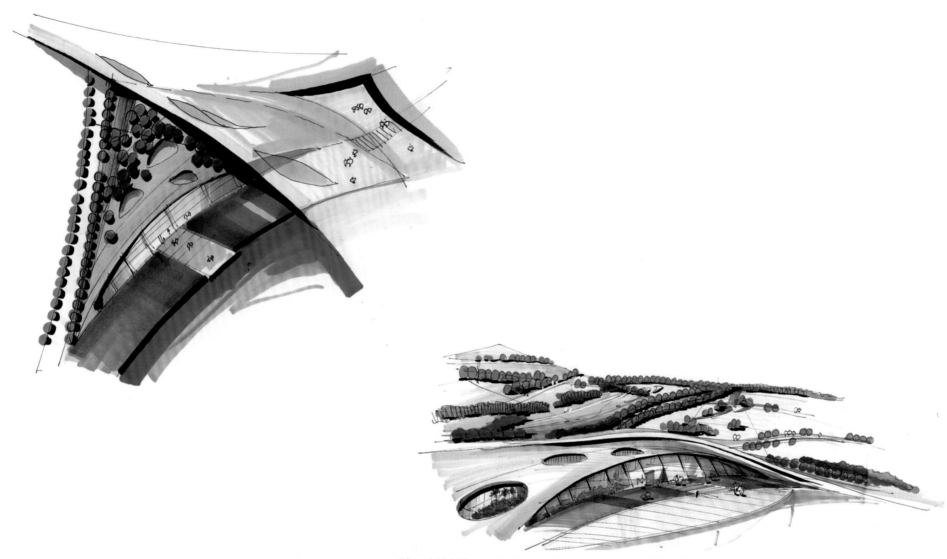

植物园温室展馆鸟瞰效果图　　Aerial perspective to the greenhouse pavillion of Botanic Garden

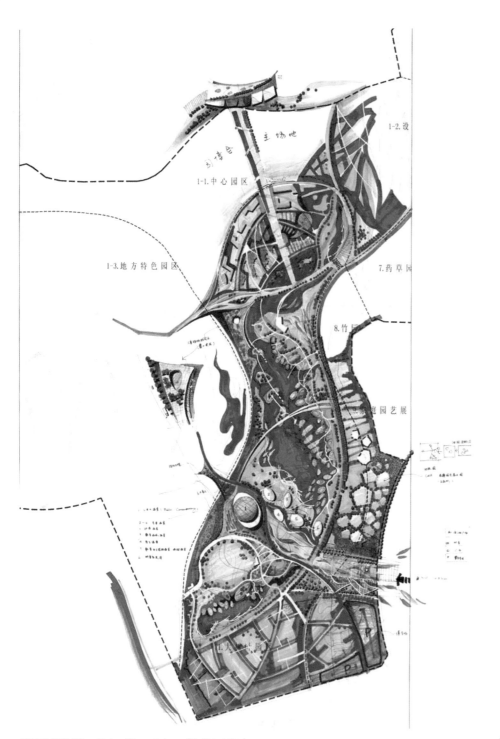

植物园中心场地设计　　Design of the central area of the Botanic Garden

节点剖面　　Node section

温室植物展厅内景透视　　Indoor perspective of the Planting Hall in the greenhouse

大温室设计表现图　　Rendering of the Big Greenhouse

植物园主山、主湖的鸟瞰图　　Aerial view of the Botanic Garden's main hill and main lake

植物园西南口　　Southwest entrance of the Botanic Garden

入口广场鸟瞰　　Aerial view of Entrance Plaza

透视竹山的林间剧场   Perspective of the Forest Theatre on the Bamboo Hill

西南入口透视,面向主湖的植物展示区   Perspective of Southwest Entrance, plant exhibition area facing the main lake

铁岭主题公园
Theme Park in Tieling

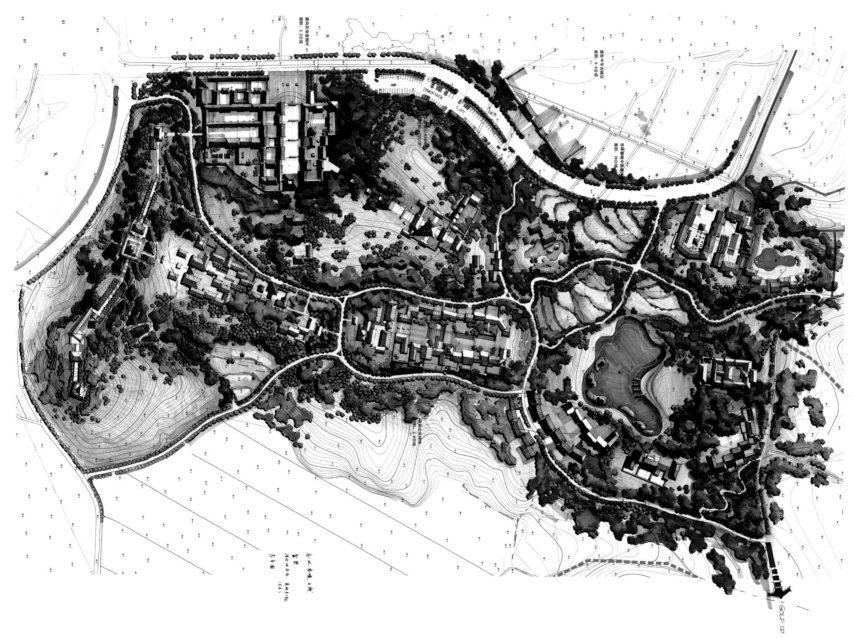

东方文化园区总平面　　The Master plan of the Oriental Culture Park

东方文化园仿古建筑底群意向图　　Intention sketch of the classic ancient type buildings in the Oriental Culture Park

东方文化园总体鸟瞰图　　Overall aerial view of the Oriental Culture Park

东方文化园中心岛建筑剖面图　Working draft of the Center Island in the Oriental Culture Park

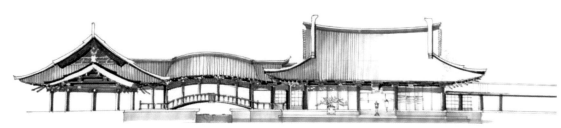

唐人文化园建筑剖面图　Working draft on the Tang Chinese Culture Park architecture design

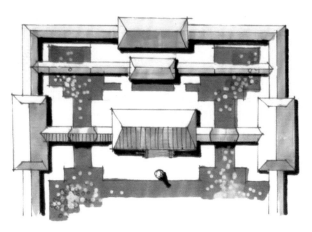

唐人文化园建筑设计草图（净土式庭园意向）　Architectural sketch of the Tang Chinese Culture Park architecture design (Option with the Pure Land garden)

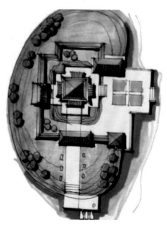

东方文化园中心岛建筑设计草图　Sketch design of the Center Island in the Oriental Culture Park

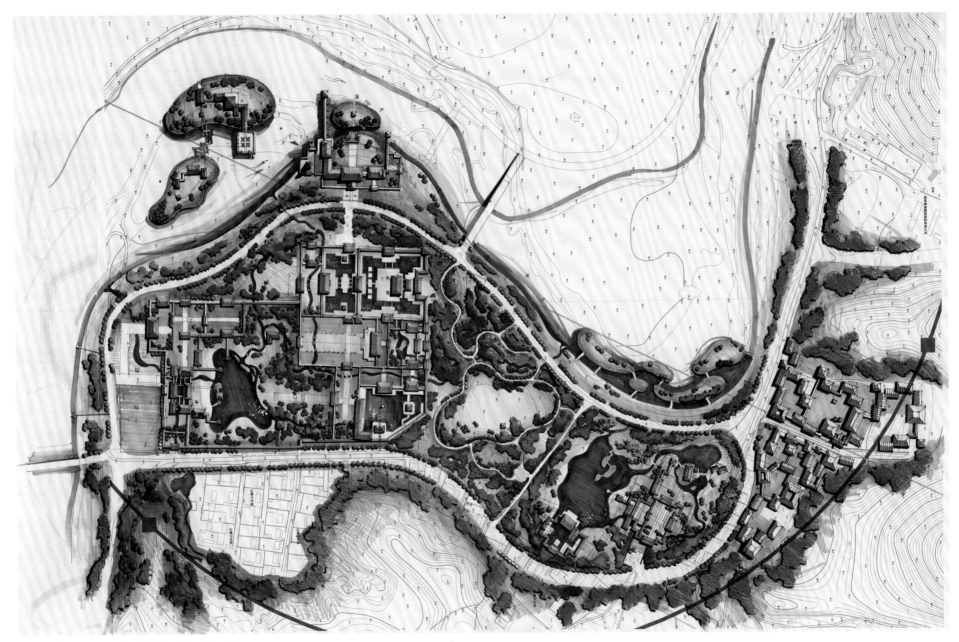

唐人文化园总平面图　　Master plan of the Tang Chinese Culture Park

山间别墅庭园意向（仿避暑山庄山区别墅布局形式）　Intention sketch of the Mountain villa garden (imitate the layout of the Imperial Summer Resort)

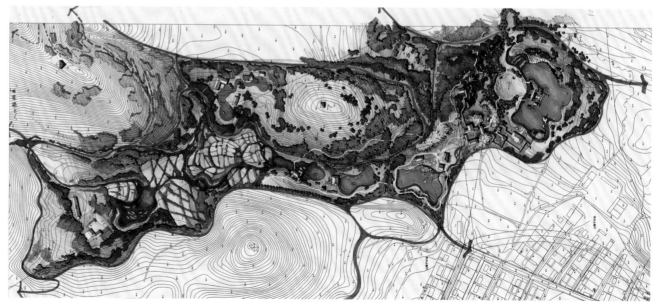

山居园（网山别墅总平面图）　Mansion in the mountain (Master plan of Wangshan Villa)

滨水休闲区效果图　Rendering of the waterfront Recreation Area

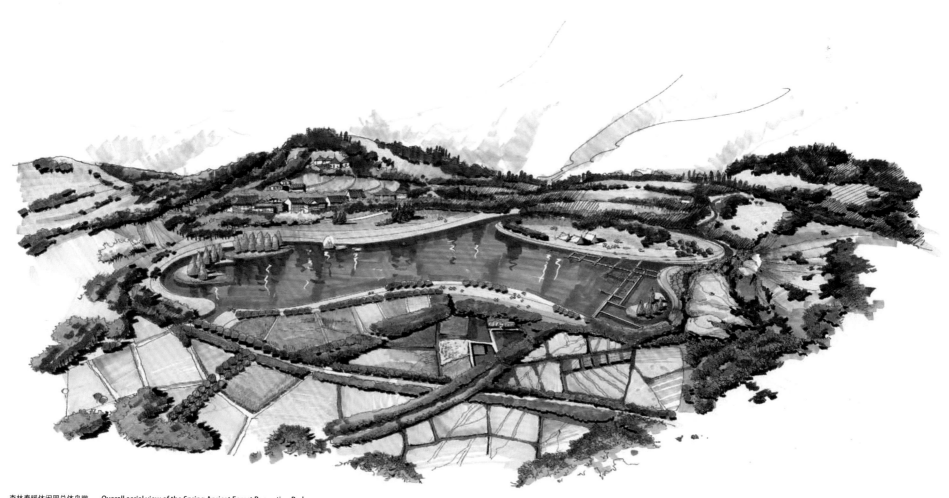

杏林春暖休闲园总体鸟瞰　　Overall aerial view of the Spring Apricot Forest Recreation Park

杏林春暖仿古建筑鸟瞰　　Aerial view of the archaized buildings in Spring Apricot Forest Garden

杏林春暖康体休闲区鸟瞰　　Aerial view of the Recreation Area in Spring Apricot Forest Garden

休闲康体中心中庭意向　　Intention of the Recreation Center's courtyard

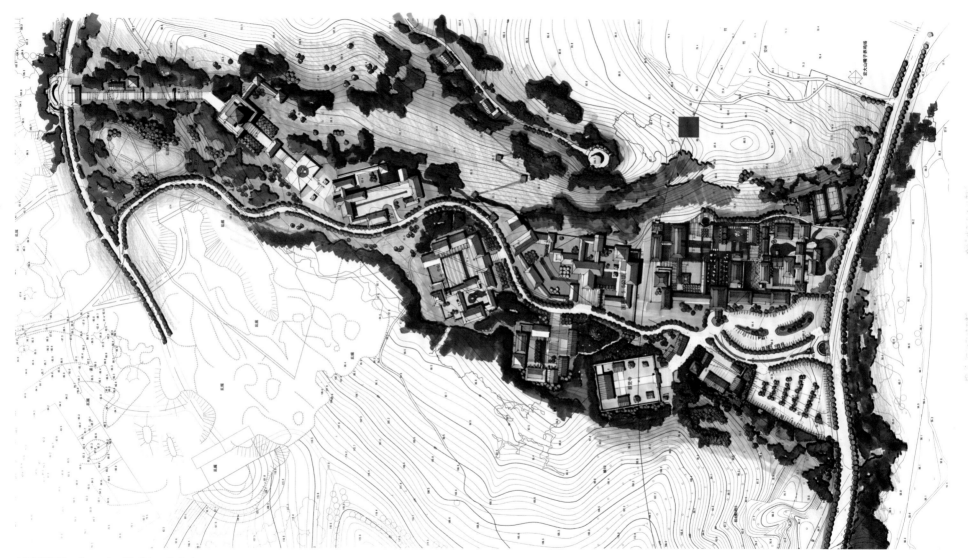

中华学园总平面　　Master plan of the Chinese Liu Garden

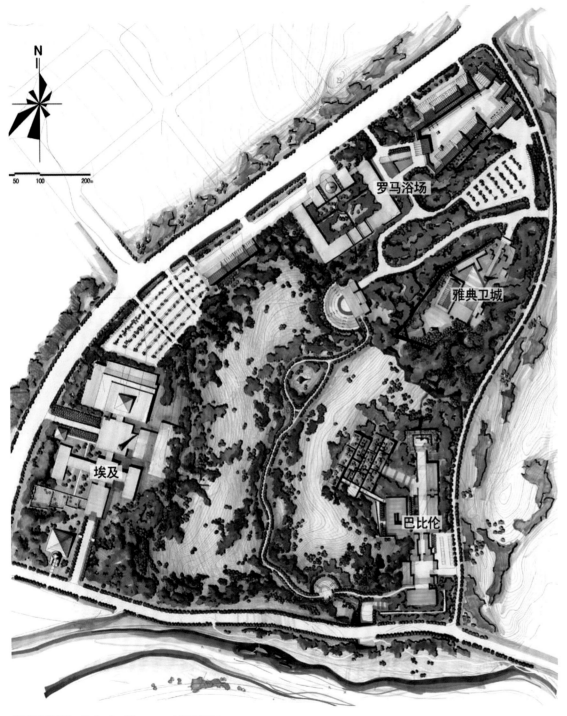

商业酒店区总平面　Master plan of the commercial/Hotel Zone

主题游乐园北区平面　Plan of the north part of the theme park

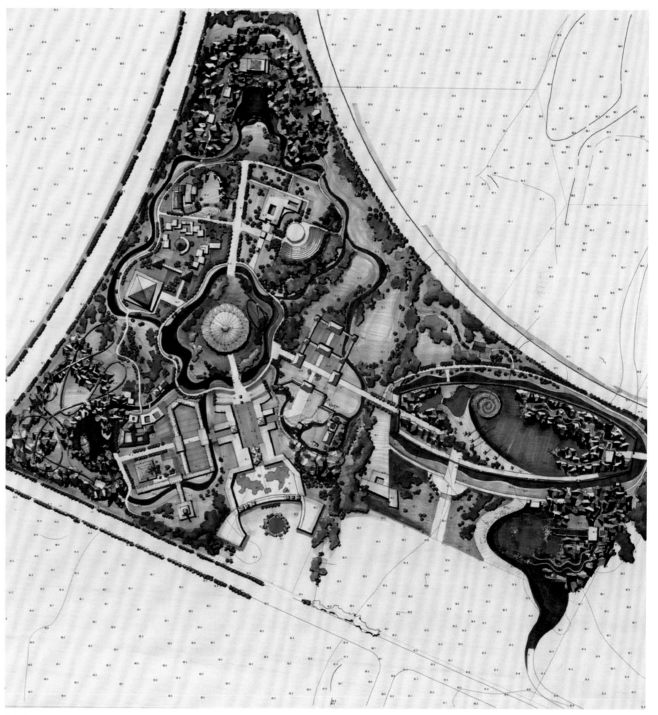

主题游乐园总平面　　Master plan of the theme park

潍坊白浪河
Bailang River in Weifang

潍坊白浪河教育城中心区鸟瞰　　Aerial view of the center zone, Bailang River Education City, Weifang

白浪河中央岛鸟瞰　　Aerial view of the Center Island of Bailang River

白浪河湿地景观带鸟瞰　　Aerial view of the wetland landscape corridor of Bailang River

白浪河"海上新城"CBD区意向图，根据日本某事务所平面规划完成的空间意向图　　Intention sketch of "New Marine City CBD zone" of Bailang River, based on the primary design of a Japanese firm

白浪河滨水公园节点　　Node of the Bailang River Waterfront Park

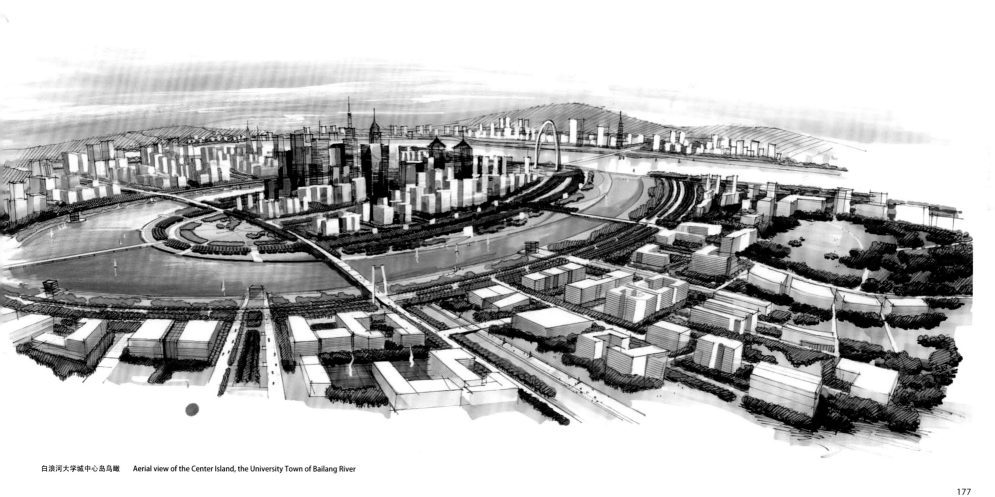

白浪河大学城中心岛鸟瞰　　Aerial view of the Center Island, the University Town of Bailang River

白浪河第二轮 CBD 区鸟瞰　　Overall aerial view of the CBD of Bailang River in the second round of planning

白浪河第二轮规划，中央区总体鸟瞰　　Overall aerial view of the Central Zone of Bailang River in the second round of planning

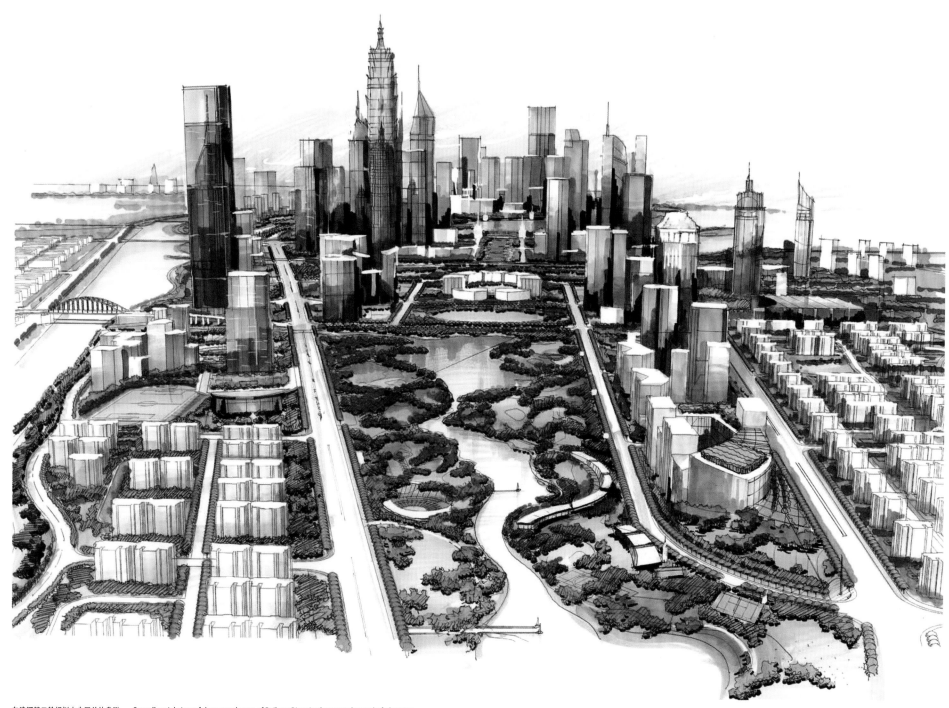

白浪河第二轮规划中央区总体鸟瞰　　Overall aerial view of the central zone of Bailang River in the second round of planning

庭院景观设计 / Landscape planning for a courtyard

注释：
以下三个项目为多年以前的设计，由于时间充裕，作品更接近于完成图纸。这类多次渲染平面图的做法与本书所论述快速平面和透视图技法有一定的差距。一并附上，供读者参考。

The following three projects were finished years ago, due to ample time, which are more close to final drawings. There are some differences between these practices, which were rendered for many times, and the quick techniques discussed for layout and perspective in this book. Now they have been attached for reference to readers.

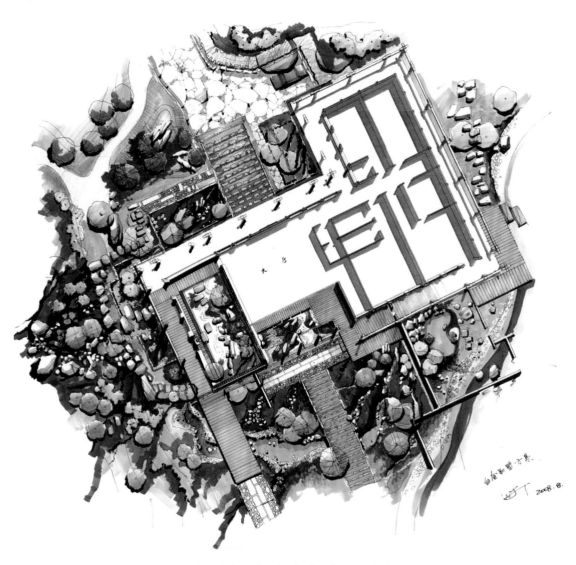

河南郑州某总统楼庭园细部平面　Layout of the garden details of the Presidential Hotel in Zhengzhou, Henan Province

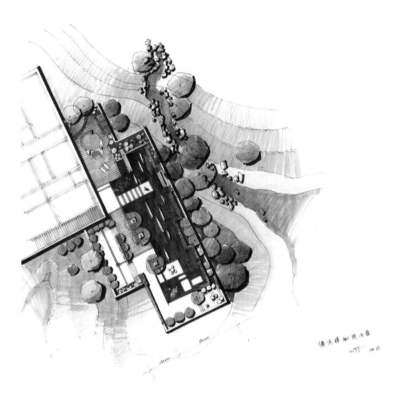

河南郑州某总统楼总平面图　　Plan of the Presidential Hotel in Zhengzhou, Henan Province

河南郑州某总统楼内庭院　　Design of the inner courtyard of the Presidential Hotel in Zhengzhou, Henan Province

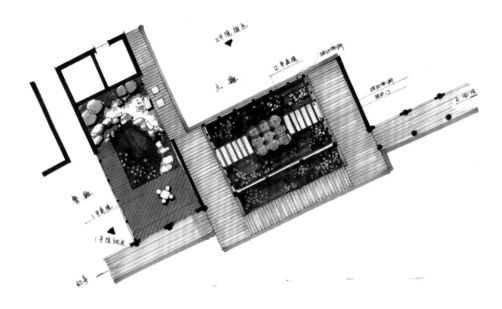

河南郑州某白金别墅庭园　　Design of a Platinum Villa Garden in Zhengzhou, Henan Province

河南郑州某总统楼庭园设计　　Design of the garden of the Presidential Hotel in Zhengzhou, Henan Province

山西朔州七里河景观规划
Landscape planning for Qilihe River in Shanxi Province

日本风格的侧庭院　Side-courtyard in Japanese-style

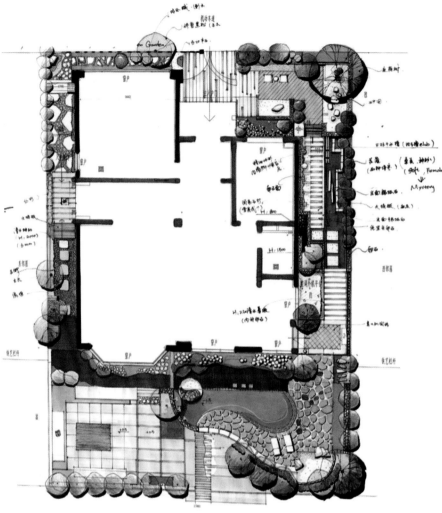

总平面　Master plan

庭院意象

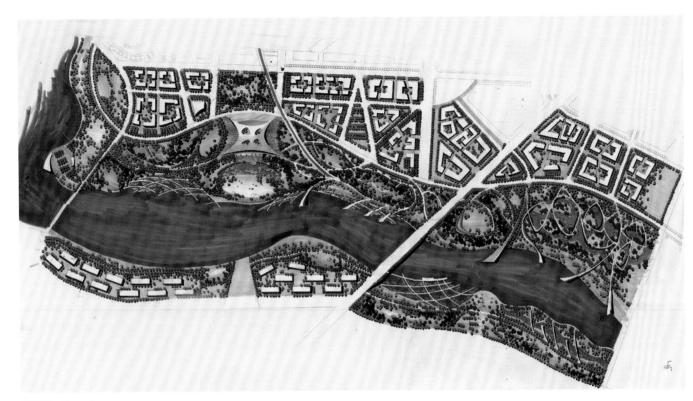

总平面图　　Master plan

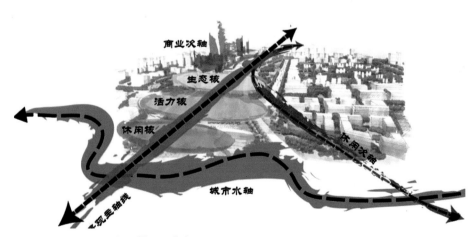

中轴线方案分析　　Program analysis of the central axis

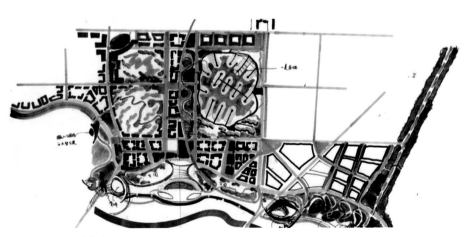

一草：平面图　　First draft: plan

滨水节点意象

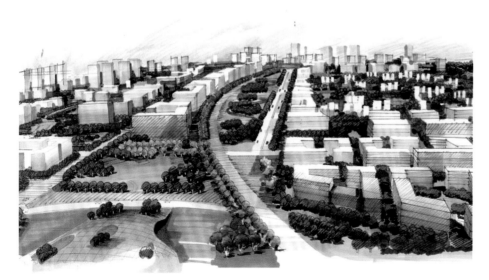

城市中轴鸟瞰

七里河城市区域总体鸟瞰

中央市民舞台——同一场地在四季的不同使用方式和意境

武汉东湖风景区规划
Planning for the East Lake Scenic Resort

武汉小南湖规划平面　　Plan of the Small South Lake, Wuhan

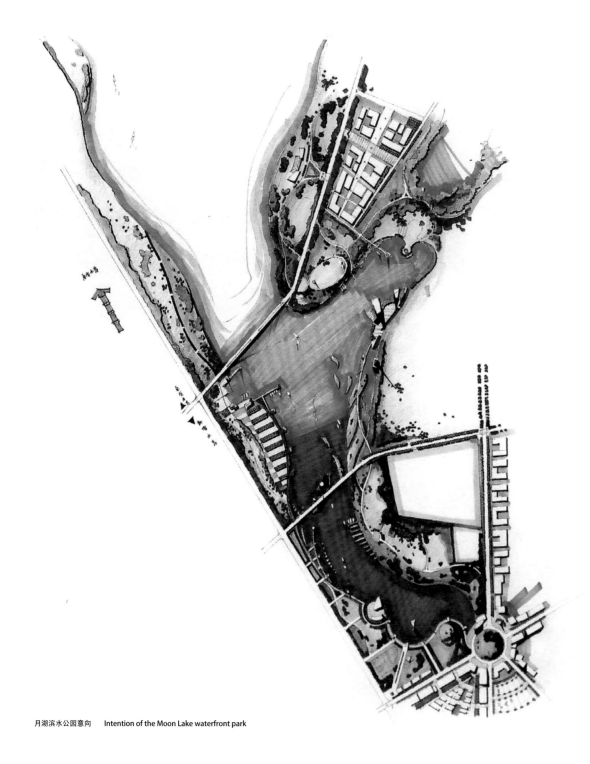

月湖滨水公园意向　　Intention of the Moon Lake waterfront park

东湖农业旅游区意向图　　Intention sketch of the East Lake Agricultural Tourism Area

北京师范大学中心景观设计
Central Plaza for Beijing Normal University

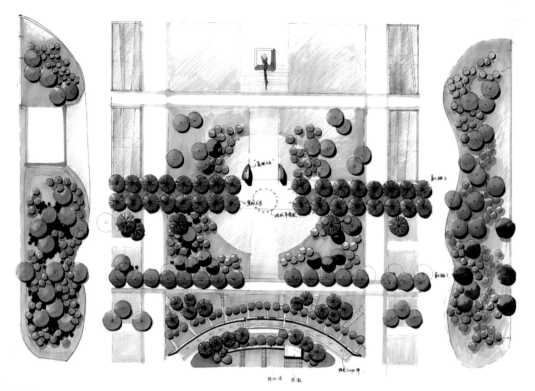

中心广场总平面图　Site Plan

标志墙　LOGO Section

中心广场透视图　　Perspective of central plaza

中心广场透视图　　Perspective of central plaza

淄博农业园景观规划
Landscape planning of the agriculture garden in Zibo

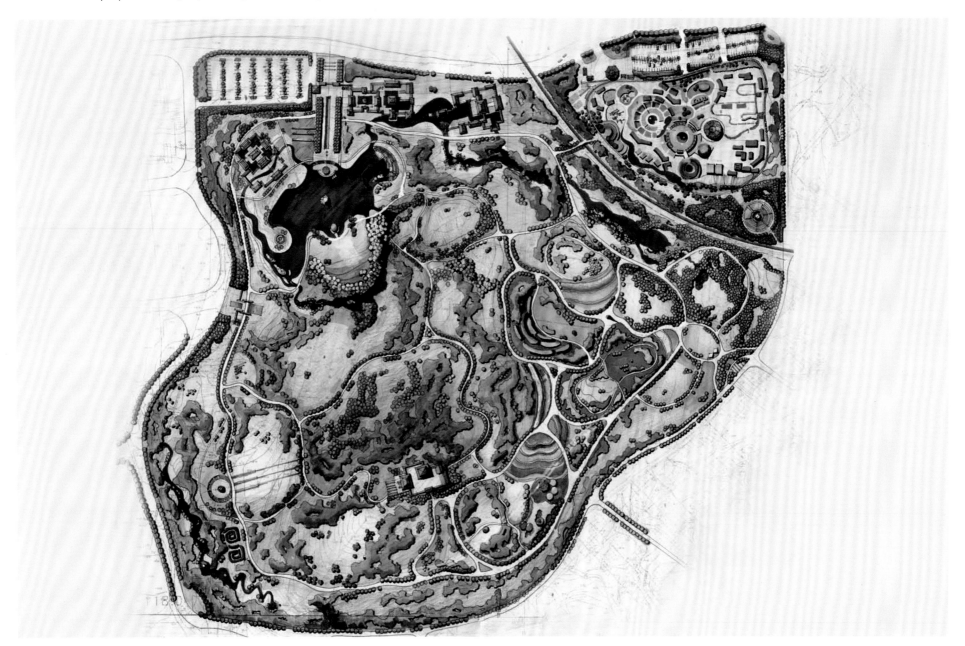

某植物园规划　　Plan of a botanic park

场地入口透视　　Perspective of the site entrance

西班牙马德里巴德维巴斯公园
Valdebebas Park in Madrid, Spain

注释：
西班牙马德里无花果公园平面图，利用场地多变的竖向高差，设计引人入胜的置水溪流和雨洪收集系统。

Plan of the Fig Garden, Madrid, Spain. Changing vertical heights on site was utilized to design attractive water streams and rainwater collection system.

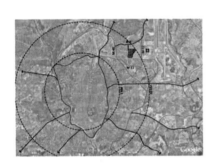
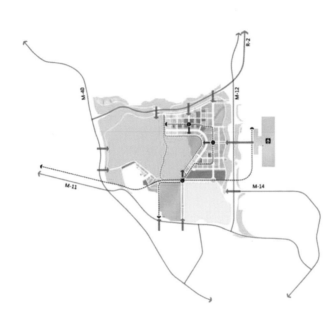
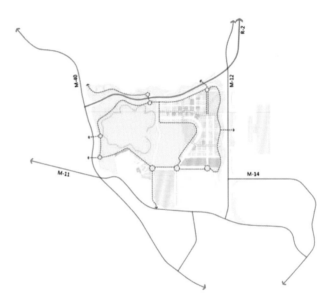

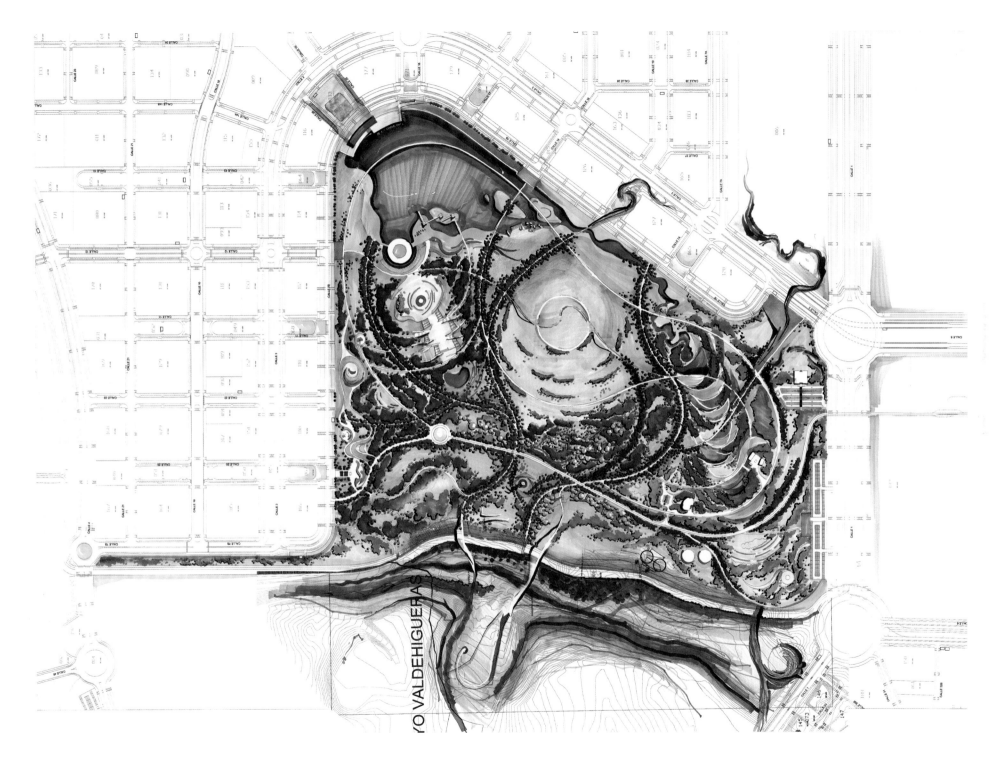

■ 景观手绘表现图赏析
Analysis of Hand Drawn Landscape

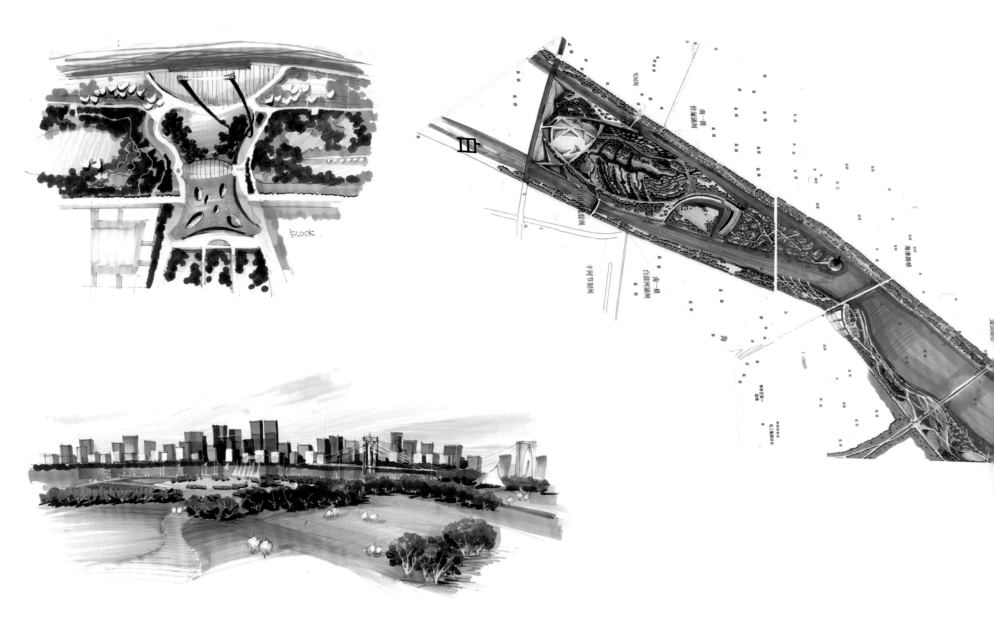

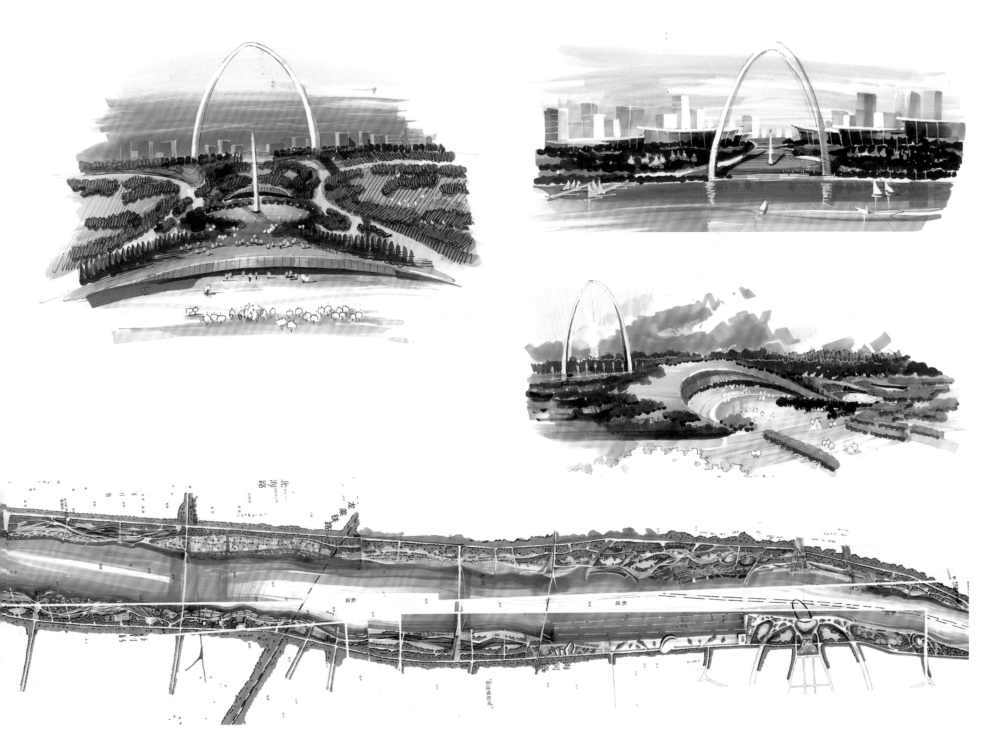

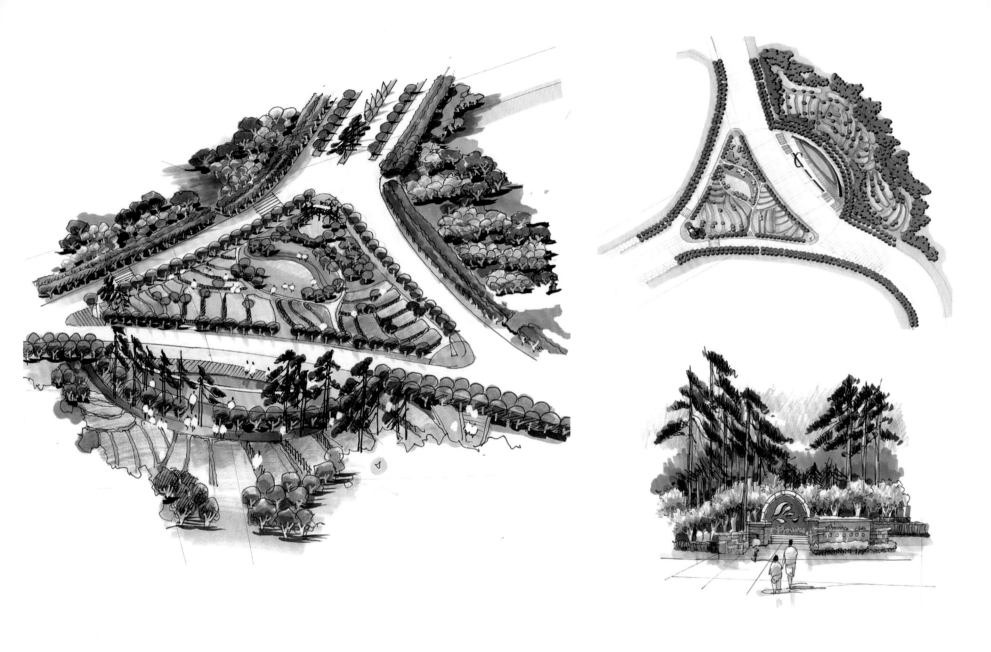

中心水花园鸟瞰一

中心水花园鸟瞰二

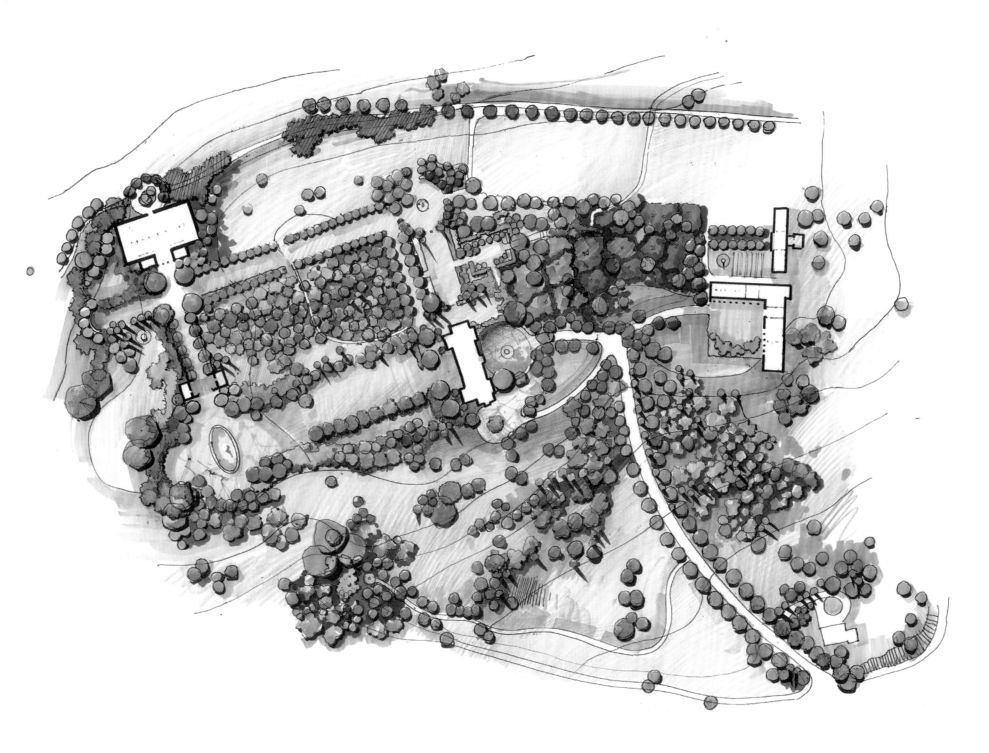

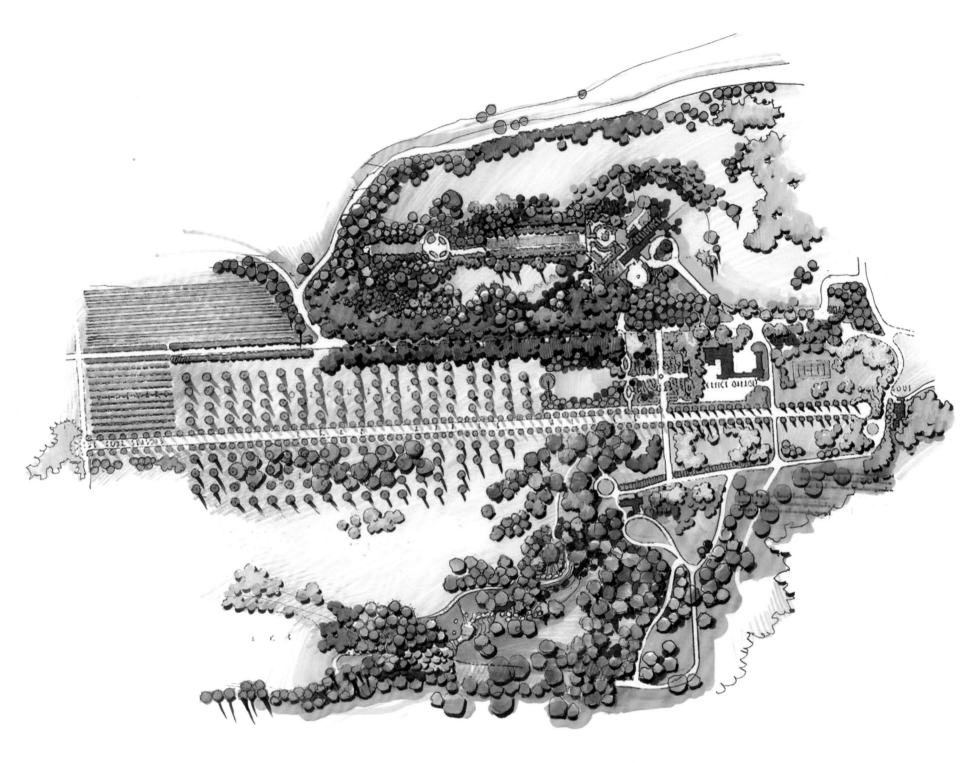